# Concise Dictionary of

# ENVIRONMENTAL ENGINEERING

## Thomas M. Pankratz

## CRC Press
Taylor & Francis Group
Boca Raton London New York

CRC Press is an imprint of the
Taylor & Francis Group, an **informa** business

CRC Press
Taylor & Francis Group
6000 Broken Sound Parkway NW, Suite 300
Boca Raton, FL 33487-2742

First issued in hardback 2017

© 1996 by Taylor & Francis Group, LLC
CRC Press is an imprint of Taylor & Francis Group, an Informa business

No claim to original U.S. Government works

ISBN 13: 978-1-138-42411-1 (hbk)
ISBN 13: 978-1-56670-212-6 (pbk)

## Library of Congress Cataloging-in-Publication Data

Pankratz, Tom M.
    Concise dictionary of environmental engineering / Thomas Pankratz.
        p.   cm.
    ISBN 1-56670-212-7
    1. Environmental sciences—Dictionary. 2. Environmental engineering—Dictionary. 3. Air—Pollution—Dictionary. 4. Water--Pollution—Dictionary. I. Title.
GE10.P36   1996
628'.03—dc20                       96-16853
                                              CIP

Library of Congress Card Number 96-16853

# Preface

There seems to be an environmental implication in almost everything we do. The field of environmental engineering has grown to the point that it overlaps the professional and private lives of everyone.

Like most technical disciplines, environmental science and engineering is undergoing an increasing degree of specialization. Industry professionals have begun to focus on specific environmental subjects and have thus become less familiar with environmental problems and solutions outside their area of expertise.

This problem is compounded by the fact that many environmentally related terms are confusing, even to a professional. Prefixes such as bio-, enviro-, hydra-, and hydro- are used so frequently that it is often difficult to tell the words apart.

This book has been written to provide industry professionals, students, and lay people with a comprehensive list of the increasing number of terms accompanying the growth of this field. More than 5200 terms, acronyms, and abbreviations have been defined from an environmental engineering perspective. Topics covered include wastewater, potable water, and industrial water treatment, seawater desalination, air pollution, incineration, hazardous waste remediation, and health-related issues.

The most unique feature of this book is the inclusion of definitions for more than 2200 commercial terms. Many of these clever, descriptive brand names for proprietary products or processes are so common that they have fallen into general use. The definitions for these terms include the company name associated with them, and the appendix contains the names, addresses, and phone and fax numbers for the companies mentioned.

During the research for this book, many other books, magazines, dictionaries, glossaries, buyer's guides, catalogs, and technical papers were reviewed to locate new terms and their definitions. Although there are too many references to list, I would like to acknowledge the help of these publications and their authors.

It was often frustrating to find hard-to-reconcile differences in definitions from one source to another, but whenever this occurred, I attempted to corroborate each definition through the use of a third source.

In addition to technically reviewing this book, John B. Tonner was especially helpful with his suggestions and research assistance, particularly during the time I lived in the Dubai, United Arab Emirates.

I would also like to thank Terrance Driscoll and Paul Gross for technically reviewing portions of the book, and the representatives of many of the companies who helped confirm product names and definitions.

I would like to acknowledge the libraries that were used to research these terms. They include the M. D. Anderson Library at the University of Houston, the Helen Hall Library in League City, Texas, the City of Houston's Central Public Library, and the library at King Fahd University of Petroleum and Mining in Dhahran, Saudi Arabia.

Much of my work on this book was done while traveling; the rest was done in the evenings and on weekends. I would never have finished without the patience and support of my wife, Julie and our children, Chad, Sarah, Michael, and Katie.

This book is dedicated to my grandmothers, Marie Ruchhoeft Felber and Catherine Mullane Pankratz, and my parents, Mike and Bette Pankratz for always encouraging me to read with a dictionary nearby.

**Tom Pankratz**
League City, Texas

# Introduction

This dictionary contains terms used in the field of environmental engineering, and the definitions provided relate to their use in an environmental context only. Environmental topics covered include wastewater collection and treatment, potable water treatment and distribution, industrial water treatment, seawater desalination, air pollution, and hazardous waste remediation.

Terms are alphabetized on a letter-by-letter basis, with spaces, hyphens, numbers, capitalization, and special characters and symbols ignored. Acronyms, abbreviations, and entries that consist of multiple words are alphabetized as if they were a single word.

In general, terms related to plumbing, "hardware," and household products have not been included. Nor have terms related to computer programs or software.

The commercial terms represent company brand names or trademarks, and have been italicized to differentiate them from the technical terms in general usage. Whenever appropriate, the use of ™ or ® has been included following the name of the entry, although a term may be a registered trademark even if it does not include either symbol. It is also possible that some of the entries listed as "trademarks" may not be registered or properly used by the manufacturers listed in connection with them.

Commercial acronyms are included if they are registered trademarks or commonly used abbreviations of company names. Nonregistered product model numbers and trademarks that are the same as the name of a company are not always included.

The company name included in the definition of a commercial term usually represents the company that manufactures that particular product or process. In some cases, the company may only market, distribute, or license the product.

In several instances, the same brand name has been listed more than once to describe different products or processes from different companies. The author is unaware of any dispute involving these cases and is simply reporting that the companies identified have used the term for the product described. In some cases, the term may be dormant or obsolete, or no longer available from the company listed.

Company addresses and phone numbers listed in the appendix were confirmed over a period of several years. It is probable that some addresses or phone/fax numbers have changed, especially with the recent telephone area code changes in many parts of the United States. Readers are cautioned that an incorrect phone number or address does not mean that a company is no longer in business.

All of the terms have been listed in good faith. A reasonable attempt has been made to confirm all definitions, and in the case of commercial terms, verify the company responsible for the listing. We apologize for any omissions or errors.

If you are aware of any changes or additions that should be included in subsequent editions, please send them to Tom Pankratz, P.O. Box 75064, Houston, Texas, 77234-5064, USA.

# A

Å   See "Angstrom."

**AA**   See "atomic absorption spectrophotometry."

**AAEE**   American Academy of Environmental Engineers.

**AAQS**   Ambient air quality standards.

*ABA-1000®*   Alumina oxide for phosphate reduction by Selecto, Inc.

*ABA-2000®*   Alumina oxide for lead and heavy metals removal by Selecto, Inc.

*ABA-8000®*   Alumina oxide for fluoride removal by Selecto, Inc.

*ABC Filter*™   Automatic backwashable cartridge filter by U.S. Filter Corp.

*Abcor®*   Ultrafiltration membrane product by Koch Membrane Systems, Inc.

*ABF*   Activated biofiltration wastewater treatment system by Infilco Degremont, Inc.

*ABF*   Traveling bridge type automatic backwashing gravity sand filter by Aqua-Aerobic Systems, Inc.

**abiotic**   The nonliving elements in the environment.

*ABJ*   Austgen Biojet Wastewater Systems.

**ABS**   (1) Acrylonitrile-butadiene-styrene, a black plastic material used in the manufacture of pipes and other components. (2) Alkyl-benzene-sulfonate, a surfactant formerly used in synthetic detergents that resisted biological breakdown.

**absolute humidity**   The amount of water vapor present in the air, measured in grams per cubic meter.

1

**absolute pressure**   The total pressure in a system, equal to the sum of the gauge pressure and atmospheric pressure.

**absolute purity water**   Water with a specific resistance of 18.3 megohm-cm at 25°C.

**absorbent**   Any substance that exhibits the properties of absorption.

**absorption**   Assimilation of molecules or other substances into the physical structure of a liquid or solid without chemical reaction.

*ABW*®   Traveling bridge type gravity sand filter by Infilco Degremont, Inc.

**AC**   See "activated carbon."

*Accelapak*®   Modular water treatment plant by Infilco Degremont, Inc.

*Accelator*®   Solids contact clarifier with primary and secondary mixing zones by Infilco Degremont, Inc.

*Accel-o-Fac*™   Sewage treatment plant design by Lake Aid Systems.

*Accelo-Biox*®   Modular wastewater treatment plant by Infilco Degremont, Inc.

*Accelo Hi-Cap*   Filter underdrain block formerly offered by Infilco Degremont, Inc.

*Access Analytical*   Former name of IDEXX Laboratories, Inc.

*Accuguard*™   Automated pH electrode cleaning and calibration module by Leeds & Northrup.

*Accu-Mag*   Electromagnetic flow meter by Wallace & Tiernan, Inc.

**accumulator**   A tank installed in a circulating water system to allow for fluctuations in flow, temperature, pressure, or other variation in operation.

*Accu-Pac*®   Polyvinyl chloride cross-corrugated media for biological wastewater treatment by Brentwood Industries, Inc.

*Accura-flo*®   Flumes for measuring flows by Hinde Engineering Co. of California, Inc.

*Accuvac*   Chemical reagents in vacuum vials for chemical analysis of fluids by Hach Company.

**ACI**   American Concrete Institute.

**acid**   (1) A substance that can react with a base to form a salt. (2) A substance that can donate a hydrogen ion or a proton.

**acid deposition**   See "acid rain."

**acidity**   The capacity of an aqueous solution to neutralize a base.

**acid rain**   Precipitation having an unusually low pH, generally attributed to the absorption of sulfur dioxide pollution in air.

*ACM*®   Thin film reverse osmosis membrane by TriSep Corp.

*Acme*   Former screening equipment manufacturer.

**acre-foot**   The volume of water that would cover a 1-acre area 1 foot deep. Equivalent to 1233.6 cubic meters.

**acrolein**   An aldehyde compound used as a microbiocide and in the manufacture of organic chemicals.

*Acro-Pac*®   Packaged seawater reverse osmosis system by Aqua-Chem, Inc.

*ACS*   Atlantes Chemical Systems, Inc.

*ACS-Plus*   High-purity chemicals for laboratory use by Hach Company.

*ACT*™   Combined aeration technologies by Aeration Industries, Inc.

*ACT-100*®   Double-wall fiberglass laminated steel underground tank by Steel Tank Association.

*Actifil®*   Packing media for biological reactors by Aeration Engineering Resources Corp.

*Actiflo®*   Drinking water treatment process by Krüger, Inc.

**activated alumina**   A partially dehydrated form of aluminum oxide frequently used as an adsorbent. Chemical formula is $Al_2O_3$.

*Activated Biofilm Method*   Fixed-film wastewater treatment system by Ralph B. Carter Co.

**activated biofilter**   Fixed-film biological wastewater treatment process with recycle of return sludge to reactor influent.

**activated carbon (AC)**   A highly adsorbent form of carbon used to remove dissolved organic matter from water and wastewater or to remove odors and toxic substances from gaseous emissions.

**activated charcoal**   See "activated carbon."

**activated sludge**   The biologically active solids in an activated sludge process wastewater treatment plant.

**activated sludge process**   A biological wastewater treatment process in which a mixture of wastewater and biologically enriched sludge is mixed and aerated to facilitate aerobic decomposition by microbes.

**activation energy**   The energy required to initiate a process or reaction.

*Activator*   Package wastewater treatment plant by Pollution Control, Inc.

*Activator III*   Oil recovery product by Sybron Chemicals, Inc.

**active life**   The period of operation of a facility that begins with initial receipt of a solid waste and ends at completion of closure activities.

**active portion** Any area of a facility where treatment, storage, or disposal operations continue to be conducted.

*Activol*™ Wastewater grease emulsifier by Probiotic Solutions.

*Acumem* Reverse osmosis products by NWW Acumem, Inc.

*Acumer*® Water treatment polymers by Rohm & Haas.

*Acutec* Gas detection system by Wallace & Tiernan, Inc.

**acute toxicity** A poisonous effect produced by a single short-term exposure that results in severe harm or death.

*AD* Dry blending and dilution system by Komax Systems, Inc.

**ADA** American Desalting Association. Formerly "NWSIA."

*Adcat*™ Oxidation catalyst systems for air pollution control by Goal Line Environmental Technologies.

*Addigest*® Package extended aeration wastewater treatment plant by Smith & Loveless, Inc.

**adiabatic lapse rate** The constant rate at which temperatures decrease as altitude increases. In a dry atmosphere the dry adiabatic lapse rate (DALR) is approximately $-1.00°C$ per 100 m rise.

*AdjustAir*® Adjustable coarse bubble air diffuser by FMC Corp., MHS Division.

*Adjust-O-Pitch* Mixing propellers with adjustable pitch blades by Walker Process Equipment Co.

**admixture** (1) A material or substance added in mixing. (2) A substance other than cement, aggregate, or water that is mixed with concrete.

*Ad-Ox* Polishing scrubber for odor abatement by Purafil, Inc.

*Adpec* Horizontal vacuum filter by Komline-Sanderson.

*Adsep*™    Chromatographic process for separating organic and inorganic compounds by U.S. Filter Corp.

*Adsolv*    Activated carbon volatile organic compound control system by RaySolv, Inc.

**adsorbate**    A material adsorbed on the surface of another.

**adsorption**    The process of transferring a substance from a liquid to the surface of a solid where it is bound by chemical or physical forces.

*Adsorption Clarifier*™    Upflow buoyant media flocculator/clarifier by Wheelabrator Engineered Systems, Inc., Microfloc Products.

*Advance*®    Chlorine gas feeder systems by Capital Controls Co., Inc.

*Advanced Fluidized Composting*™    A combined biological and chemical sludge treatment process by ERM Group.

**advanced oxidation processes (AOPs)**    Processes using a combination of disinfectants, such as ozone and hydrogen peroxide, to mineralize toxic organic compounds to nontoxic form.

**advanced secondary treatment**    Secondary wastewater treatment with enhanced solids separation.

**advanced wastewater treatment (AWT)**    Treatment processes designed to remove pollutants that are not adequately removed by conventional secondary treatment processes.

*Advent*™    Package water treatment plant by Infilco Degremont, Inc.

**aeolian deposit**    Soil deposited by the wind.

*Aeralator*    Iron and manganese removal unit by General Filter Co.

**aeration**    The addition of air or oxygen to water or wastewater, usually by mechanical means, to increase dissolved oxygen levels and maintain aerobic conditions.

*Aeration Panel*™   Fine-bubble membrane diffuser panel by Parkson Corp.

**aerator**   A device used to introduce air or oxygen into water or wastewater.

*Aercor*   Aeration and packing products by Aeration Engineering Resources Corp.

*Aer-Degritter*   Aerated grit removal system by FMC Corp., MHS Division.

*AerFlare*   Air diffuser by Walker Process Equipment Co.

*Aergrid*™   Floor grid aeration system by Aeration Technologies, Inc.

*Aermax*™   Fine-pore aeration diffuser by Aeration Technologies, Inc.

*Aero-Accelator*®   Circular, packaged activated sludge treatment plant by Infilco Degremont, Inc.

**aerobe**   An organism that requires oxygen for respiration.

**aerobic**   Condition characterized by the presence of oxygen.

**aerobic digestion**   Sludge stabilization process in which aerobic biological reactions destroy biologically degraded organic components of sludge.

*Aeroburn*   Wastewater treatment plant by Walker Process Equipment Co.

*Aerocleve*   Former manufacturer whose product line is now offered by Chemineer, Inc.

*Aeroductor*   Aerated grit removal system by Lakeside Equipment Corp.

*Aero-Filter*   Rotary distributor by Lakeside Equipment Corp.

*Aer-O-Flo*   Wastewater treatment equipment product line by Purestream, Inc.

*Aero-Max*   Tubular membrane diffuser by Aeration Research Company.

*Aero-Mod*   Modular aeration and clarification units by Aero-Mod, Inc.

*Aeropure*   Activated carbon vapor filtration system by American Norit Company, Inc.

*AeroScrub*   Flue gas scrubber by Aeropulse, Inc.

*Aerosep*®   Multistage aerosol separation system by Kimre, Inc.

**aerosols**   Liquid particles small enough to remain suspended and dispersed in air, or another gas, for a period of time.

*Aero-Surf*   Air driven rotating biological contactor by Envirex, Inc.

*Aerotherm*   In-vessel composting system by Compost Systems Co.

*AerResearch*   Aeration diffuser product line by Aeration Research Company.

*Aershear*™   Coarse-bubble diffuser by Aeration Technologies, Inc.

*Aertec*™   Air diffuser product line by Aeration Technologies, Inc.

*Aertube*™   Static tube aerators by Aeration Technologies, Inc.

**afforestation**   The process of establishing a forest where one did not previously exist.

**AFO**   Air fail open.

**AFPA**   American Forest and Paper Association.

*AfterBlend*   Output booster for chemical feed system by Stranco, Inc.

**afterburner**   A device used to reduce air emissions by incinerating organic matter in a gas stream.

**aftercondenser**   A condenser installed as the last stage of an evaporator venting system to minimize atmospheric steam discharge.

*AFX*™   Ozone instrumentation products by IN USA, Inc.

**agar**  A gelatinous substance extracted from a red algae, commonly used as a medium for laboratory cultivation of bacteria.

**agar plate**  A circular glass plate, containing a nutrient, used to culture microorganisms.

**Agent Orange**  A mixture of organochlorine herbicides, sometimes containing dioxins, used as a defoliant during the Vietnam War.

**agglomerate**  To gather fine particles into a larger mass.

*Agidisc*  Disc filter with integrated agitation system by Eimco Process Equipment Co.

*Agisac*  Sock-type screening sack by Hydro-Aerobics, Inc. (U.S.) and Copa Group (U.K.).

**AGMA**  American Gear Manufacturers Association.

**agrichemical**  Any inorganic, artificial, or manufactured chemical substances used in agricultural processes, usually as fertilizers, herbicides, and pesticides.

**agronomy**  Branch of agriculture that deals with the raising of crops and the care of the soil.

**A-horizon**  Topsoil, or the uppermost layer of soil containing the highest accumulation of mineral and organic matter.

**AHS**  See "aquatic humic substances."

**AIChE**  American Institute of Chemical Engineers.

**AIDS**  Acquired immune deficiency syndrome.

**air**  The mixture of gases, primarily oxygen and nitrogen, that surrounds the earth and forms its atmosphere.

*Airamic®*  Air/gas diffuser by Ferro Corp.

*AiRanger*  Tank level measurement system by Milltronics, Inc.

*Airbeam™*  Aluminum aeration basin cover by Enviroquip, Inc.

**air-bound**  Obstruction of water flow in a pipeline or pump due to the entrapment of air.

*Airbrush*™   Rotor aerator by United Industries, Inc.

*Airco*   Former name of BOC Gases.

*Air Comb*®   Coarse-bubble diffuser by Amwell, Inc.

*Aircushion*   Flotation clarifier by Wilfley Weber, Inc.

**air diffuser**   A device designed to transfer atmospheric oxygen into a liquid.

*Aire-O2*®   Propeller aspirator aerator by Aeration Industries, Inc.

*Air Grid*   Sand filter air scour system by Roberts Filter Manufacturing Co.

*Air-Grit*   Aerated grit removal system by Walker Process Equipment Co.

*AirLance*™   In-vessel composting technology by CBI Walker, Inc. (licensee) and American Bio Tech, Inc. (licensor).

**airlift**   A device for pumping liquid by injecting air at the bottom of a riser pipe submerged in the liquid to be pumped.

*Air Mix*   Pulsing bed filter surface cleaning process by Zimpro Environmental, Inc.

*Airmizer*   Air diffuser by EnviroQuip International Corp.

*Air-O-Lator*®   Floating aerator by Air-O-Lator Corp.

*AiroPump*   Airlift pump by Walker Process Equipment Co.

*AirOXAL*®   Pure oxygen process by Liquid Air.

**air pollutant**   Airborne gases, liquids, or solids that may be hazardous to animal or plant life.

**air pollution**   The presence in the atmosphere of any airborne gases, liquids, or solids that may be hazardous to animal or plant life.

**air quality related value (AQRV)**   A value referring to the reduction in visibility that may be caused by a new air emission.

*AirRide*   Density controlling system for compost in-feed by Waste Solutions.

**air scour**   The agitation of granular filter media with air during the filter backwash cycle.

*Air Seal*   Coarse-bubble diffuser by Jet, Inc.

*Airsep*   Aerated grit collector system by Aerators, Inc.

**air stripper**   The process of removing volatile and semi-volatile contaminants from liquid; air and liquid are passed countercurrently through a packed tower.

*AirTainer*™   Tank cover by NuTech Environmental Corp.

**air-to-cloth ratio**   The ratio of the volumetric flow rate of a gas to be filtered to the fabric area of the filter.

*Airvac*®   Vacuum sewage collection system by Airvac, Inc.

**AISC**   American Institute of Steel Construction.

**AISI**   American Iron and Steel Institute.

*Akta Klor*   Sodium chlorite solution by Rio Linda Chemical Co.

**alabaster**   A compact, fine-grained gypsum material.

*Albrivap*   High-temperature additive for evaporators by Albright & Wilson.

**alcohol**   A class of compounds containing the hydroxyl group OH.

**aldehyde**   A class of organic compounds containing a CHO group, including formaldehyde and acetylaldehyde.

**algae**   Primitive, free-floating, plant-like aquatic organisms. The singular form is "alga."

*AlgaeMonitor*   On-line fluorometer to monitor relative algae levels in a water system by Turner Designs.

*Algae Sweep Automation*   Automated clarifier algae sweep system by Ford Hall Co., Inc.

**algal blooms** Rapidly reproducing floating colonies of algae that may cover streams, lakes, and reservoirs, creating nuisance conditions.

*AlgaSORB®* Ion exchange medium for heavy metal removal by Bio-Recovery Systems, Inc.

**algicide** Any substance used to kill algae. Also "algaecide."

**aliphatic compounds** Organic compounds with carbon atoms arranged in a straight or branched chain, rather than a ring.

**aliquot** The amount of a sample used for analysis.

**alkali** A substance with highly basic properties.

**alkali metals** The elements lithium, sodium, potassium, rubidium, and cesium.

**alkaline** Water containing sufficient amounts of alkalinity to raise the pH above 7.0.

**alkaline soil** Soil with a pH greater than 7.0.

**alkalinity** The ability of a water to neutralize an acid due to the presence of carbonate, bicarbonate, and hydroxide ions.

*Alkalinity First™* Sodium bicarbonate by Church & Dwight Co., Inc.

*Alka-Pro®* Process control system for biological wastewater treatment systems by Davis Water & Waste Industries, Inc.

*Allison* Internally fed rotating drum screen by KRC (Hewitt) Inc.

**alluvial soil** Soil formed of material that was carried by flowing water before being deposited.

**alpha factor** The ratio of oxygen transfer coefficients for water and wastewater at the same temperature and pressure; used in the sizing of aeration equipment.

**alpha ray** A stream of particles emitting from the nucleus of a helium atom undergoing disintegration.

*Altech*   Continuous emissions monitoring product line by Wheelabrator Clean Air Systems, Inc.

**alternative energy**   Energy obtained from sources other than traditional fossil fuels or nuclear energy, and which are usually renewable and nonpolluting. Alternative energy sources include solar energy, wave power, geothermal power, and biomass fuels.

**alum**   Common name for aluminum sulfate, frequently used as a coagulant in water and wastewater treatment. Chemical formula is $Al_2(SO_4)_3 \cdot 14H_2O$.

*Alumadome*   Self-supporting aluminum covers for circular tanks by Conservatek Industries, Inc.

*Alumavault*   Self-supporting aluminum covers for rectangular tanks by Conservatek Industries, Inc.

*Alumdum*   Porous diffuser dome by Aeration Engineering Research Corp.

**alumina**   A form of aluminum oxide. Chemical formula is $Al_2O_3$.

**aluminum**   A lightweight, nonferrous metal with good corrosion resistance and electrical and thermal conductivity.

**aluminum sulfate**   See "alum."

**alum sludge**   Sludge resulting from treatment process where alum is used as a coagulant.

*AlumStor*   Modular liquid storage tank and feed system by ModuTank, Inc.

*Amberjet*™   Ion exchange resins by Rohm & Haas Co.

*Ambersomb*®   Carbonaceous adsorbent for volatile organic compound removal by Rohm & Haas Co.

**ambient air quality**   A general term used to describe the condition of the outdoor air.

*Amcec*   Volatile organic compound control products by Wheelabrator Clean Air Systems, Inc.

**amendment** Organic material, such as wood chips or sawdust, added to sludge in a composting operation to promote uniform air flow.

*Amerfloc* Polyelectrolyte used to enhance liquid/solid separation by Ashland Chemical, Drew Division.

*America Norit* Former name of Norit Americas, Inc.

*American Well Works* Former name of Amwell, Inc.

*Ameroid* Polyelectrolyte used to enhance liquid/solid separation by Ashland Chemical, Drew Division.

*Ames Crosta* Sludge thickening system by Biwater Treatment Ltd.

**ammonia** A compound of hydrogen and nitrogen that occurs extensively in nature. Chemical formula is $NH_3$.

**ammonia nitrogen** The quantity of elemental nitrogen present in the form of ammonia.

**ammonification** Bacterial decomposition of organic nitrogen to ammonia.

**ammonium ion** A form of ammonia found in solution, the ion $NH_4^+$.

**amoeba** A single-celled protozoan microbe. Also "ameba."

**amoebiasis** See "amoebic dysentery." Also "amebiases."

**amoebic dysentery** A form of dysentery caused by a protozoan parasite, usually resulting from poor sanitary conditions and transmitted by contaminated food or water. Also "amebic dysentery."

**amorphous** Noncrystalline, having no shape or form.

**amperometric titrator** Titration device containing an internal indicator or electrometric device to show when the reactions are complete.

**amphoteric** Capable of reacting in water either as a weak acid or weak base.

**AMSA** Association of Metropolitan Sewerage Agencies.

**anabatic wind**   A localized wind that flows up valley or mountainous slopes, usually in the afternoon, caused by the replacement of cool valley air with the warmer air above it.

**anaerobe**   An organism that can thrive in the absence of oxygen.

**anaerobic**   Condition characterized by the absence of oxygen.

**anaerobic digestion**   Sludge stabilization process where the organic material in biological sludges is converted to methane and carbon dioxide in an airtight reactor.

*Anaerobic Selector Process*   Biological system for phosphorus and biochemical oxygen demand removal by Davis Water & Waste Industries, Inc.

*Analite*   Portable turbidimeters by Advanced Polymer Systems.

*An-CAT®*   Polymer processing control unit by Norchem Industries.

*Anco*   Batch mixers by Enviropax, Inc.

**anemometer**   An instrument used to measure the force or velocity of wind.

**aneroid barometer**   An instrument used to measure atmospheric pressure that operates on the movement of a thin metal plate rather than the rise and fall of mercury.

**angle of repose**   The maximum angle that the inclined surface of a loosely divided material can make with the horizontal.

*Anglian Engineering Services*   Former equipment manufacturer now part of Rosewater Engineering Ltd.

**Angstrom (Å)**   A unit of measure equivalent to one ten-thousandth of a micron.

**anhydride**   A chemical compound derived by the elimination of water.

**anhydrite**   See "calcium sulfate."

**anhydrous**   A compound that does not contain water.

**anion**   A negatively charged ion that migrates to the anode when an electrical potential is applied to a solution.

**anionic polymer**   A polyelectrolyte with a net negative electrical charge.

*Anitron*   Biological fluidized bed wastewater treatment system by Krüger, Inc.

*ANM*™   Nanofiltration softening membrane by TriSep Corp.

*Annubar*®   Mass-flow monitoring system by Dieterich Standard.

**anode**   The positive electrode where current leaves an electrolytic solution.

**anodic protection**   Electrochemical corrosion protection achieved through the use of an anode having a higher electrode potential than the metal to be protected.

**anoxic**   Condition characterized by the absence of free oxygen.

*ANOX-R*   Advanced treatment industrial wastewater treatment system by Davis Water & Waste Industries, Inc.

**ANSI**   American National Standards Institute.

**anthracite**   A hard black coal containing a high percentage of fixed carbon and a low percentage of volatile matter, and which burns with little or no smoke.

*Anthrafilt*®   Filter anthracite by Unifilt Corp.

**anthropogenic compounds**   Compounds created by human beings, often relatively resistant to biodegradation.

**antifoam agent**   A surface active agent used to reduce or prevent foaming.

**antifoulant**   An additive or dispersant that prevents fouling and/or the formation of scale.

**antigen**  A substance capable of stimulating an immune response.

**antiknock additive**  A compound, usually tetraethyl lead, added to gasoline to minimize engine preignition and its accompanying knocking and pinging. Pollution from the release of such compounds in auto emissions led to the introduction of unleaded gasoline.

**antioxidant**  A substance that slows down or prevents oxidation of another substance.

**antiscalant**  An additive that prevents the formation of inorganic scale.

*A/O®*  Biological system for phosphorus and biochemical oxygen demand removal by Krüger, Inc.

**AOC**  See "assimilable organic carbon."

**AOPs**  See "advanced oxidation processes."

**APCA**  American Pollution Control Association.

**APHA**  American Public Health Association.

**API**  American Petroleum Institute.

**API gravity**  An index inversely related to specific gravity used to identify liquid hydrocarbons.

**API separator**  Rectangular basin in which wastewater flows horizontally while free oil rises and is skimmed from the surface.

*APOVAC®*  Antipollution vacuum system for solvent recovery by Rosenmund, Inc.

**apparent color**  The color in water caused by the presence of suspended solids.

**approach velocity**  The average water velocity of fluid in a channel upstream of a screen or other obstruction.

**APWA**  American Public Works Association.

**AQRV**  See "air quality related value."

*Aqua-4*™  Surface water treatment plant by Smith & Loveless, Inc.

*AquaABF*   Traveling bridge type automatic backwash package filter by Aqua-Aerobic Systems, Inc.

*Aqua Bear*   Medium-density foam pipeline cleaners by Girard Industries.

*Aquabelt®*   Gravity belt thickener by Ashbrook Corp. (U.S.) Simon-Hartley Ltd. (U.K.).

*AquaCalc*   Open channel flow computer by JBS Instruments.

*Aqua-Carb*   Activated carbon by Wheelabrator Clean Water Systems, Inc.

*Aqua-Cat®*   Sulfide conversion process to control odors by Wheelabrator Clean Air Systems, Inc.

*Aquaclaire™*   Wastewater treatment systems by DAS International, Inc.

*Aqua Criss Cross*   Medium-density coated foam pipeline cleaners by Girard Industries.

**aquaculture**   The managed production of fish or shellfish in a pond or lagoon.

*AquaDDM*   Direct drive mixers by Aqua-Aerobic Systems, Inc.

*Aquadene®*   Corrosion and scale control products by Stiles-Kem.

*AquaDisk*   Woven cloth tertiary filter by Aqua-Aerobic Systems, Inc.

*Aquafeed®*   Reverse osmosis antiscalants by B.F. Goodrich Co.

*Aqua-Fer™*   Well water treatment plant designed for iron removal by Smith & Loveless, Inc.

*Aqua Guard™*   Self-cleaning bar screen by Parkson Corp.

*Aqua-Jet*   Direct-drive aerators by Aqua-Aerobic Systems, Inc.

*Aqua-Lator®*   High-speed floating aerator by Aerators, Inc.

*Aqualenc* Aluminum chlorosulfate by Rhone-Poulenc Basic Chemicals Co.

*AquaLift®* Screw pump by Parkson Corp.

*Aquamag®* Magnesium hydroxide by Premier Services Corp.

*Aquamite®* Electrodialysis water treatment systems by Ionics, Inc.

*Aqua Pigs™* Polyurethane foam pipeline cleaners by Girard Industries.

*Aquaport®* Seawater desalination system by Ambient Technologies, Inc.

*Aquaray®* Ultraviolet disinfection system by Infilco Degremont, Inc.

*Aquaritrol®* Coagulant control system using a programmable controller by Wheelabrator Engineered Systems, Inc., Microfloc Products.

*Aquarius®* Modular water treatment plant by Wheelabrator Engineered Systems, Inc., Microfloc Products.

*AquaSBR* Sequencing batch reactor by Aqua-Aerobic Systems, Inc.

*Aquascan* On-line volatile organic compound monitor by Sentex Systems, Inc.

*Aqua-Scrub™* Powdered activated carbon adsorption system by Wheelabrator Clean Water Systems, Inc.

*Aqua-Sensor* Control system for water softener regeneration by Culligan International Corp.

*Aquashade®* Aquatic plant growth control by Applied Biochemists, Inc.

*Aqua-Shear* Mixer by Flow Process Technology, Inc.

*Aquasorb®* Carbon adsorption treatment systems by Hadley Industries.

*Aquasource®* Membrane system by Infilco Degremont, Inc.

*Aquaspir®*   Shaftless dewatering screw by Andritz-Ruthner, Inc.

*Aquastore®*   Storage tanks by A.O. Smith Harvestore Products, Inc.

*Aqua Swab*   Soft polyurethane pipeline cleaners by Girard Industries.

*Aquatair*   Packaged biological wastewater treatment plant by BCA Industrial Controls, Ltd.

*Aquatech Systems*   Former name of Aqualytics, Inc.

**aquatic humic substances (AHS)**   Humic substances in true solution that exhibit colloidal properties.

*Aquatreat™*   Sequencing batch reactor by EnviroSystems Supply, Inc.

*Aqua-Trim™*   Tray-type air stripper by Delta Cooling Towers, Inc.

*Aqua UV™*   Ultraviolet water disinfection systems by Trojan Technologies, Inc.

*Aquavap*   Vapor compression type evaporator by Licon, Inc.

*Aqua-View*   Particle measuring system by Particle Measuring Systems, Inc.

*Aquaward®*   Tablet feeder disinfection system by Eltech International Corp.

*Aquazur V®*   Rapid sand gravity filter by Infilco Degremont, Inc.

**aqueduct**   A conduit for carrying running water.

**aqueous chlorine**   Term used to describe chlorine or chlorine compounds dissolved in water, often mistakenly called "liquid chlorine."

**aqueous solution**   A solution in which water is the solvent.

**aquiclude**   A low-permeability underground rock formation that absorbs water slowly but will not allow its free passage.

**aquifer** A subsurface geological formation containing a large quantity of water.

**aquifuge** An underground layer of impermeable rock that will not allow the free passage of groundwater.

*Aquox*™ Potassium permanganate by Nalon Chemical, distributed by American International Chemical, Inc.

**arable** Land capable of being farmed.

**ARAR** Applicable or relevant and appropriate requirements. Cleanup standards, control standards, and other substantive environmental protection requirements, criteria, and limitations promulgated under federal, state, and local laws.

**Archimedes' principle** The principle of buoyancy that states that the force on a submerged body acts vertically upward through the center of gravity of the displaced fluid and is equal to the weight of the fluid displaced.

**Archimedes' screw** See "screw pump."

*Arc Screen*™ Self-cleaning curved bar screen by Infilco Degremont, Inc.

*Arcticaer* High-speed surface aerator with submersible motor by Aerators, Inc.

*ARI* Sulfur recovery and odor control product line by Wheelabrator Clean Air Systems, Inc.

**arithmetic mean** The sum of a set of observations divided by the number of observations.

*Arm & Hammer*® Sodium bicarbonate by Church & Dwight Co., Inc.

*Armco* Former sluice gate manufacturer acquired by Hydro Gate Corp.

*Arna*® Ultraviolet disinfection system by Arlat, Inc.

**aromatics** A group of hydrocarbon compounds, including benzene, containing a closed ring structure.

**array** A group of solar collection devices arranged in a suitable pattern to efficiently collect solar energy.

*Arro-Care*®   Reverse osmosis maintenance and support services by U.S. Filter Corp.

*Arro-Cleaning*®   Membrane care services by U.S. Filter Corp.

*Arrowhead*®   Water treatment product line by U.S. Filter Corp.

**arroyo**   A stream or watercourse that is often dry.

*ARS*™   Air regulated siphon by John Meunier, Inc.

**arsenic**   A naturally occurring element that is toxic to humans at very low levels. Chemical formula is As.

**artesian water**   Bottled water from a well that taps a confined aquifer located above the normal water table.

**artesian well**   A well with sufficient pressure to produce water without pumping.

*Arus Andritz*   Former name of Andritz-Ruthner, Inc.

**ASA**   Algae sweep automation system by Ford Hall Co., Inc.

**asbestos**   A mineral fiber that does not conduct heat or electricity. It was formerly in wide use in the building industry for thermal insulation, soundproofing, roofing, and electrical insulation, and has since been positively identified as a carcinogen.

**asbestos cement pipe**   Pipe manufactured of a mixture of asbestos fiber and Portland cement.

**asbestos-containing material**   Construction materials that contain more than 1% asbestos.

**asbestosis**   A chronic lung disease caused by exposure to or inhalation of asbestos fibers.

**ASCE**   American Society of Civil Engineers.

**aseptic**   The state of being free of pathogenic organisms.

**ash**   The nonvolatile inorganic solids that remain after incineration.

*Ashaire*   Aerator product line by Aerators, Inc.

*Ashbrook-Simon-Hartley*   Former name of Ashbrook Corp.

*Ashfix*™   A process to stabilize heavy metals in sludge and ash by Ashland Chemical, Drew Division.

**ASHRAE**   American Society of Heating, Refrigeration and Air Conditioning Engineers.

**ASME**   American Society of Mechanical Engineers.

*Aspergillus fumigatus*   Airborne fungi that may result from composting operations and may cause human ear, lung, and sinus infections.

**asphalt-rubber**   A mixture of ground rubber and bituminous concrete used as a pavement interlayer to reduce stress and prevent cracking.

**aspirating aerator**   Aeration device that uses a motor-driven propeller to draw atmospheric air into the turbulence caused by the propeller to form small bubbles.

**aspirator**   A hydraulic device that creates a negative pressure by forcing liquid through a restriction and increasing the velocity head.

**ASQC**   American Society for Quality Control.

**assimilable organic carbon (AOC)**   The portion of dissolved organic carbon that is easily used by microbes as a carbon source.

**assimilative capacity**   The ability of a water body to receive wastewater and toxic materials without deleterious effects on aquatic life or the humans who consume the water.

**AST**   Aboveground storage tanks.

**ASTM**   American Society for Testing and Materials.

**asymmetric membrane**   Membranes that are not reversible and can only desalinate efficiently in one direction.

*AT*®   Ozone generator by Ozonia North America.

**ATAD**   Autothermal thermophilic aerobic digestion process.

*Atara* Former manufacturer of digester gas mixing equipment whose product line was acquired by Infilco Degremont, Inc.

*ATD*™ Autothermal aerobic sludge digestion system by CBI Walker, Inc.

**atm** See "atmosphere."

**atmometer** An instrument used to measure the evaporative capacity of the air.

**atmosphere** (1) The gaseous region that surrounds the earth. (2) A unit of pressure equal to 1.0333 kg/sq cm, or 14.7 pounds per square inch. Abbreviated "atm."

**atmospheric corrosion** Corrosion resulting from exposure to the atmosphere.

**atmospheric pressure** The force exerted by the weight of the atmosphere above the point of measurement.

*Atochem* Former name of Elf Atochem North America.

**atom** The smallest unit of an element that retains the characteristics of that element.

*Atomerator* Pressure aerator by General Filter Co.

**atomic absorption spectrophotometry (AAS)** A highly sensitive instrumental technique for measuring trace quantities of elements in water.

**atomize** To divide a liquid into extremely fine particles.

**atomizer** An instrument through which a liquid is sprayed to produce a fine mist.

*ATP*™ Aerobic thermophilic process sludge treatment system by CBI Walker, Inc.

**ATSDR** Agency for Toxic Substances and Disease Registry.

**attached growth process** See "fixed film process."

**attainment area** A geographic area in which the levels of a criterion air pollutant meet the health-based national ambient air quality standard for that pollutant.

*Auger Monster*®   Modular wastewater screen/grinder headworks system by JWC Environmental.

*Auto5*™   Stack sampling system by Greasby.

*AutoBelt*   Rotary vacuum filter formerly offered by Walker Process Equipment Co.

**autoclave**   A device that sterilizes materials by exposure to pressurized steam.

*Auto-Cleanse*   Self-cleaning pump station with differential pressure activated flush valve by ITT Flygt.

*Auto-Dox*   Controlled aeration weir by Purestream, Inc.

**autogenous combustion**   Burning that occurs when the heat of combustion of a wet organic material or sludge is sufficient to vaporize the water and maintain combustion without auxiliary fuel.

**autogenous temperature**   Equilibrium temperature in sludge combustion where the heat input from the fuel equals the heat losses and combustion is self-supporting.

*Auto-Jet*®   Pressure leaf filter by U.S. Filter Corp.

*Autojust*   Automatic feedwell gate controller for circular sludge collectors by Aerators, Inc.

**auto-oxidation**   A self-induced oxidation process.

*Auto-Pulse*™   Tubular backpulse filter by U.S. Filter Corp.

*Auto-Rake*®   Reciprocating rake bar screen by Franklin Miller.

*Auto-Retreat*   Automatic bar screen control system by Infilco Degremont, Inc.

*AutoSDI*™   Portable, computer-based silt density index instrument by King Lee Technologies.

*Auto-Shell*™   Granular media filter by U.S. Filter Corp.

*Auto-Shok*™   Tube-type vertical pressure leaf filter by U.S. Filter Corp.

*Auto-Skimmer*™    Skimmer used to remove floating hydrocarbons from water wells by R.E. Wright Associates, Inc.

*AutoTherm*™    Aerobic thermophilic digestion system by CBI Walker, Inc.

**autothermal thermophilic aerobic digestion**    A biological digestion system that converts soluble organics to lower-energy forms through anaerobic, fermentative, and aerobic processes at thermophilic temperatures.

**autothermic combustion**    See "autogenous combustion."

*Autotravel*    Traveling bridge sludge collectors by Simon-Hartley, Ltd.

*Autotrol*    Former manufacturer of rotating biological contactors whose product line was acquired by Envirex, Inc.

**autotroph**    Organism that derives its cell carbon from carbon dioxide.

*Auto-Vac*    Rotary drum filter by Alar Engineering Corp.

*AVGF®*    Automatic valveless gravity filter by U.S. Filter Corp.

**AWS**    American Welding Society.

**AWT**    See "advanced wastewater treatment."

**AWWA**    American Water Works Association.

**axial flow**    The flow of fluid in the same direction as the axis of symmetry of a tank or basin.

**axial flow pump**    A type of centrifugal pump in which fluid flow remains parallel to the flow path and develops most of its head by the lifting action of the vanes.

**azeotropic**    A liquid of two or more substances that behaves like a single substance in that the vapor produced by partial evaporation has the same composition as the liquid.

*Aztec*    Water monitors by Capital Controls Co.

# B

**B-10 life**   The rated life defining the number of revolutions that 90% of a group of identical bearings will complete before first evidence of failure develops.

**BAC**   Biologically active carbon.

**bacilli**   Rod-shaped bacteria.

**backfill**   The material used to refill a ditch or excavation, or the process of refilling.

**backflow**   Flow reversal in a water distribution system that may result in contamination due to a cross-connection.

**backflow prevention device**   Device used to prevent cross-connection or backflow of nonpotable water into a potable water system.

**background concentration**   The general level of air pollutants in a region with all local sources of pollution ignored.

**background contamination**   Contamination introduced accidentally into dilution waters, reagents, rinse water, or solvents that can be confused with constituents in the sample being analyzed.

**background soil pH**   The pH of the soil prior to the addition of substances that alter the hydrogen ion concentration.

**back pressure**   Pressure due to a force operating in a direction opposite to that required.

**backsiphonage**   A backflow of water of questionable quality that results from a negative pressure within the water distribution system.

**backwash** A high-rate reversal of flow for the purpose of cleaning or removing solids from a filter bed or screening medium.

**backwash rate** The flow rate used during filter backwash, when the direction of flow through the filter is reversed for cleaning.

**BACT** Best available control technology.

**bacteria** Microbes that decompose and stabilize organic matter in wastewater.

**baffle** A plate used to provide even distribution, or to prevent short-circuiting or vortexing of flow entering a tank or vessel.

*Baffleflow* Oil removal tank with permeable baffles to prevent short-circuiting by Walker Process Equipment Co.

**bagasse** Crushed sugar cane or sugar beet refuse from sugar making.

**baghouse** An air emissions control device that uses a fabric or glass fiber filter to remove airborne particulates from a gas stream.

*Bakflo* Barge mounted oil skimmer by Vikoma International.

**balefill** A land disposal site where solid waste material is compacted and baled prior to disposal.

**baler** A machine used to compress and bind solid recyclable materials such as cardboard or paper.

**ballast water** Water used in a ship's hold for stabilization, often requiring treatment as an oily wastewater.

**ball valve** A valve utilizing a rotating ball with a hole through it that allows straight-through flow in the open position.

**bandscreen** See "traveling water screen."

**bank sand** Sand excavated from a natural deposit, usually not suitable for use in filter processing or grading.

**bar**   A unit of pressure equal to 0.9869 atmospheres, $10^6$ dyn/cm$^2$, and 14.5 lb/in$^2$.

*Bardenpho*$^{sm}$   Biological wastewater treatment process for removal of nitrogen and phosphorus by Eimco Process Equipment Co.

*Barminutor*®   Combination bar screen and comminuting device by Yeomans Chicago Corp.

**barnacles**   A marine crustacean with a calcareous shell that attaches itself to submerged objects.

**barometer**   An instrument used to measure atmospheric pressure.

**barometric condenser**   A condenser in which vapor is condensed by direct contact with water.

**barometric damper**   A pivoting plate used to regulate the amount of air entering a duct or flue to maintain a constant draft within an incinerator.

**barometric leg**   (1) A condensate discharge line submerged below the liquid level of an atmospheric tank. (2) A gravity tailpipe from a vacuum barometric condenser.

**barometric pressure**   Ambient or local pressure surrounding a gauge, evaporator shell, vent pipe, etc.

**barrel (bbl)**   42 U.S. gallons.

**barried landscape water renovation system (BLWRS)**   A wastewater treatment and denitrification system where wastewater is applied to the top of a mound of soil overlaying a water barrier and microbes oxidize soluble organics as the water percolates through the soil.

*Barry Rake*   Trash rake by Cross Machine, Inc.

**bar screen**   Screening device using mechanically operated rakes to remove solids retained on a stationary bar rack.

*Bartlett-Snow*™   Rotary calciner for soil reclamation by ABB Raymond.

**base** (1) A substance that can accept a proton. (2) A substance that can react with an acid to form a salt. (3) An alkaline substance.

**baseline** An sample used as comparative reference point when conducting further tests or calculations.

**basicity factor** Factor used to determine neutralization capabilities of alkaline reagents used to treat acidic wastes.

**basket centrifuge** Batch-type centrifuge where sludge is introduced into a vertically mounted spinning basket and separation occurs as centrifugal force drives the solids to the wall of the basket.

**BAT** See "best available technology."

*Batch-Master* Bottom discharge basket centrifuge by Ketema, Inc.

*Batch Master* Package wastewater treatment system by Wastewater Treatment Systems, Inc.

*Batch-Miser* Horizontal plate filter by Ketema, Inc.

*Batch-O-Matic* Bottom discharge basket centrifuge by Ketema, Inc.

**batch process** A noncontinuous treatment process in which a discrete quantity or batch of liquid is treated or produced at one time.

**batch reactor** A reactor where the contents are completely mixed and flow neither enters nor leaves the reactor vessel.

**BATEA** Best available technology economically available.

**battery limit** The boundary limits of equipment, or a process unit that defines interconnecting points for electrical piping or wiring.

*Bauer®* Screening equipment product line acquired by Andritz-Ruthner, Inc.

**bauxite**   Ore containing alumina monohydrate or alumina trihydrate, which is the principal raw material for alumina production.

**bbl**   See "barrel."

**BCF**   See "bioconcentration factor."

*BCL Screen*   Back-cleaned bar screen by Jones and Attwood, Inc.

**BDAT**   See "best demonstrated available technology."

**BDOC**   See "biodegradable dissolved organic carbon."

**beachwell**   A shallow intake well making use of beach sand and structure as a filter medium.

*Bead Mover*™   Ion exchange resin and filter media loading pump by IX Services Co.

*Bead Thief*™   Ion exchange resin core sampler by IX Services Co.

**Beaufort scale**   A numerical scale of wind force on which a Beaufort force 0 wind is calm and a force 12 wind indicates hurricane force with winds in excess of 120 km/hr (75 mph).

**bed depth**   The depth of filter media or ion exchange resin contained in a vessel.

**bedrock**   Solid rock encountered below the mantle of loose rock and soil on the earth's surface.

**beggiatoa**   Filamentous microbe, commonly associated with sludge bulking, that results from low dissolved oxygen levels and/or high sulfide levels.

*Bekomat*®   Microprocessor-driven condensate trap by BEKO Condensate Systems Corp.

*Bekosplit*   Separation process for condensate emulsions by BEKO Condensate Systems Corp.

*Belclene*   Scale control additive by FMC Corp., Process Additives Division.

*Belcor*   Organic corrosion control by FMC Corp., Process Additives Division.

*Belgard*®   Antiscalant for seawater evaporators by FMC Corp., Process Additives Division.

*Belite*®   Antifoaming agent by FMC Corp., Process Additives Division.

*Bellacide*   Algicide by FMC Corp., Process Additives Division.

*BelloZon*   Chlorine dioxide generator by ProMinent Fluid Controls, Inc.

*Beloit-Passavant*   Company acquired by Zimpro Environmental, Inc.

*Belspere*   Chemical dispersant by FMC Corp., Process Additives Division.

**belt conveyor**   A device used to transport material, consisting of an endless belt that revolves around head and tail pulleys.

**belt filter press**   See "belt press."

**belt press**   A sludge dewatering device utilizing two fabric belts revolving over a series of rollers to squeeze water from the sludge.

**belt thickener**   Mechanical sludge processing device that uses a revolving horizontal filter belt to prethicken sludge prior to dewatering and/or disposal.

**bench test**   A small-scale test or study used to determine whether a technology is suitable for a particular application.

**beneficial organism**   A pollinating insect, pest predator, parasite, pathogen, or other biological control agent that functions naturally or as part of an integrated pest management program to control another pest.

**benthal oxygen demand**   The oxygen demand exerted by the organic mud and sludge deposits on the bottom of a river or stream.

**benthic**   Relating to the bottom environment of a water body.

**benthos**   Microbes and other organisms living on the bottom of a water body.

*BentoLiner*™   Clay composite liner by SLT North America, Inc.

*Bentomat*™   Geotextile-bentonite liner by Colloid Environmental Technologies Co.

**bentonite**   Colloidal clay-like mineral that can be used as a coagulant aid in water treatment systems. Also used as a landfill liner because of its limited permeability.

**benzene**   An aromatic hydrocarbon used as a solvent; it has carcinogenic properties and is often characterized by its ring structure. Chemical formula is $C_6H_6$.

**berm**   A horizontal, earthen ridge or bank.

**Bernoulli's equation**   Energy equation commonly used to calculate head pressure; it considers velocity head, static head, and elevation.

**best available technology (BAT)**   The best technology, treatment techniques, or other means available after considering field, rather than solely laboratory, conditions.

**best demonstrated available technology (BDAT)**   A technology demonstrated in full-scale commercial operation to have statistically better performance than other technologies.

**best management practice (BMP)**   The schedules of activities, methods, measures, and other management practices to prevent pollution of waters and facilitate compliance with applicable regulations.

**beta radiation**   Radiation emitted from an atomic nucleus during radiation.

*BFI*®   Trademark of Browning-Ferris Industries, Inc.

*B-Gon*™   Mist eliminator by Kimre, Inc.

*BGS2*   Sludge dryer/pelletizer formerly offered by Wheelabrator Clean Water Systems, Inc., Bio Gro Division.

*BHN Probiotic*   Lagoon sludge oxidation product line by Bio Huma Netics, Inc.

**B-horizon**   The intermediate soil layer, usually having a high clay content, where minerals and other particles washed down from the A-horizon accumulate.

**bhp**   Brake horsepower. The power developed by an engine as measured by a dynamometer applied to the shaft or flywheel.

*Bibo*   Dewatering and drainage pumps by ITT Flygt.

**bicarbonate**   A chemical compound containing an $HCO_3$ group.

**bicarbonate alkalinity**   Alkalinity caused by bicarbonate ions.

*Bi-Chem*®   Blend of selectively adapted bacterial cultures for wastewater treatment by Sybron Chemicals, Inc.

*BIF*®   Product group of Leeds & Northrup, a unit of General Signal.

*Bifad*   Fill and draw activated sludge plant developed by Ardern and Lockett and offered by Biwater Treatment Ltd.

**biflow filter**   Granular media filter characterized by water flow from both top and bottom to a collector located in the center of the filter bed.

**bilharzia**   Waterborne disease also known as "schistosomiasis."

**binary fission**   Asexual reproduction in some microbes where the parent organism splits into two independent organisms.

**bioaccumulative**   A characteristic of a chemical whose rate of intake into a living organism is greater than its rate of excretion or metabolism.

*Bio-Activation*   Combination activated sludge and trickling filter wastewater treatment system by Amwell, Inc.

**bioassay**   An analytical method that considers a change in biological activity as a means of analyzing a material's response to biological treatment or an environment.

**biobrick**   A building brick made of kiln-dried municipal wastewater sludge.

*Biocarb*™   Activated carbon by Wheelabrator Clean Water Systems, Inc.

*Biocarbone*®   Biological wastewater treatment process using an immersed fixed-bed filter by Biwater Treatment Ltd.

**biochemical oxidation**   Oxidative reactions brought about by biological activity which result in chemical combination of oxygen with organic matter.

**biochemical oxygen demand (BOD)**   A standard measure of wastewater strength that quantifies the mg/L of oxygen consumed in a stated period of time, usually five days.

*Biocidal*™   Sodium hypochlorite system by Scienco/FAST Systems.

**biocide**   A chemical used to inhibit or control the population of troublesome microbes.

*Bio-Clarifier*   Secondary clarifier with rotating sludge scoop for use with package rotating biological contactor by Envirex, Inc.

*Bioclean*   Reverse osmosis biocide by Argo Scientific.

*Bioclere*™   Packaged wastewater treatment plants by Ekofinn Bioclere.

**bioconcentration**   The net increase in concentration of a substance that results from the uptake or absorption of the substance directly from the water and onto aquatic organisms.

**bioconcentration factor (BCF)**   The accumulation of chemicals that live in contaminated environments equal to the quotient of the concentration of a substance in aquatic organisms divided by the concentration in the water during the same time period.

**biocontactor**   A unit process such as an aeration basin, trickling filter, rotating biological contactor, or digester where microbes degrade/transform organic matter.

**bioconversion**   The conversion of organic waste products into an energy resource through the action of microbes.

*Biocube*™   Aerobic biofilter for airborne odors and volatile organic compounds by EG&G Biofiltration.

*Bio-D*®   Bioremediation nutrients by Medina Products.

**biodegradable**   Term used to describe organic matter that can undergo biological decomposition.

**biodegradable dissolved organic carbon (BDOC)**   The portion of total organic carbon that is easily degraded by microbes.

*BIOdek*®   Synthetic media for fixed-film wastewater treatment reactors by Munters.

*Bio-Denipho*   Biological phosphorus removal process by Krüger, Inc.

*Biodenit*   Biological denitrification process using immersed fixed-bed filter by Biwater Treatment Ltd.

*Bio-Denitro*   Biological nitrogen removal process by Krüger, Inc.

**biodiversity**   An environment where multiple organisms coexist.

*BioDoc*®   Rotary distributor for trickling filters by Wes-Tech Engineering, Inc.

*Bio-Drum*   Rotating drum containing biological filter media for wastewater treatment by Ralph B. Carter Co.

*Bio-Energizer*   Lagoon sludge oxidation system by Probiotics Solutions division of Bio Huma Netics, Inc.

**biofeasibility**   A bioremediation feasibility study done to determine the applicability and potential success of a bioremediation technique or procedure for a given site.

**biofilm**   An accumulation of microbial growth.

**biofilter**   See "biological filter."

*Bio\*Fix*®   Alkaline stabilization process for biosolids by Wheelabrator Clean Water Systems, Inc., Bio Gro Division.

*Bioflush*   Trash rake by E. Beaudrey & Co.

*Biofor*™   Biological fixed-film wastewater treatment system by Infilco Degremont, Inc.

**biofoul**   Presence and growth of organic matter in a water system.

**biogas**   The gases produced by the anaerobic decomposition of organic matter.

**biogenesis**   The theory that living organisms arise only from other living organisms.

*Bio Genesis*™   Microbial formulation to reduce wastewater odors by Bio Huma Netics, Inc.

*Bioglas*   Rigid, open-cell foam silica biological oxidation media by the former Bioglas Corp.

*Bioglas Alpha*   Package fixed-film wastewater treatment plant by the former Bioglas Corp.

*Bio Gro*   Bio Gro Division of Wheelabrator Clean Water Systems, Inc.

*Bio Jet-7*   Organic solution of seven strains of live, nontoxic bacteria by Jet, Inc.

*Bioken*   Former name of Cuno Separations Systems Division.

**biokinetics**   The branch of science that pertains to the study of living organisms.

*Biolac*®   Extended aeration waste treatment process by Parkson Corp.

**Biolift™**   Waste activated sludge thickening system by Humboldt Decanter, Inc.

**Bio\*Lime®**   Agricultural liming agent by Wheelabrator Clean Water Systems, Inc., Bio Gro Division.

**Biologic™**   Nutrient supplement for wastewater treatment facilities by SciCorp Systems, Inc.

**biological filter**   A bed of sand, stone, or other media through which wastewater flows that depends on biological action for its effectiveness.

**biomass**   The mass of biological material contained in a system.

**Bio Max™**   Floating piping system by Environmental Dynamics, Inc.

**biome**   A biological community or ecosystem characterized by a specific habitat and climate such as a tropical rainforest or a desert.

**Biomizer**   Continuously sequencing reactor process by Environmental Dynamics, Inc.

**Bio-Module**   Package wastewater treatment plant using rotating biological contactors by Envirex, Inc.

**BioMonitor™**   Automated on-line biochemical oxygen demand analyzer by Anatel Corp.

**biomonitoring**   The use of living organisms to test water quality at a site further downstream.

**Bio-Net®**   Rotating biological contactor by NSW Corp.

**BIONOx™**   Submersible aerator/mixer by Framco Environmental Technologies.

**Bionutre**   Nutrient removal process by Envirex, Inc.

**Bio-Nutri™**   Nitrogen and phosphorus removal process by Smith & Loveless, Inc.

**Bio-Ox™**   Bioreactor for biological treatment of wastewater by SRE, Inc.

**bio-oxidation**   See "biochemical oxidation."

*Bio-Pac*   Package trickling filter plant by FMC Corp., MHS Division.

*Bio-Pac*®   Trickling filter media by NSW Corp.

*Biopaq*®   Upflow anaerobic sludge blanket process used for treatment of high-strength wastes by CBI Walker, Inc. (U.S. licensee), Biwater Treatment Ltd. (U.K. licensee), and Paques B.V. (licensor).

*Bio-Pure*   Water reclamation treatment plant by Aqua-Clear Technologies Corp.

biopure water   Water that is sterile, pyrogen free, and has a total solids content of less than 1 mg/L.

*Bio-Reel*™   Fixed-film wastewater treatment system using coiled, corrugated tubing by Schreiber Corp.

bioremediation   Application of the natural ability of microbes to use waste materials in their metabolic processes and convert them into harmless endproducts.

*Bio-S*®   Bioremediation surfactant by Medina Products.

*Bio/Scent*   Liquid odor neutralizer by Hinsilblon, Ltd.

*Bioscrub*   Rotating biological contactor odor and volatile organic compound treatment system by CMS Group, Inc.

*Bioscrubbers*™   Biological based system for removal of odors by WRc Process Engineering.

*Bio-Separator*   Floating flow diversion baffle for lagoons by ThermaFab, Inc.

*Bio-Sock*   Fabric sock used to introduce bacterial cultures into a flow by Sybron Chemicals, Inc.

*Biosock*™   Biological culture application system by Sybron Chemicals, Inc.

biosolids   Primarily organic sludges or byproducts of wastewater treatment that can be beneficially recycled.

biosorption process   See "contact stabilization process."

biosphere   The mass of living organisms found in a thin belt at the earth's surface.

*BioSpiral*    Rotating biological contactor formerly offered by Walker Process Equipment Co.

*Biostart*™    Liquid microbial concentrate by Advanced Microbial Systems, Inc.

*Biostyr*®    Upflow mixed-media reactor process for removal of nitrogen and suspended solids by Krüger, Inc.

*Bio-Surf*    Rotating biological contactor process by Envirex, Inc.

**biota**    All living organisms within a system.

*Biotac*™    Bioremediation system that delivers bacteria to wet wells by Davis Industries, Process Division.

*Biothane*®    Anaerobic wastewater treatment process by Biothane Corp.

*Bioton*®    Biological volatile organic compound and odor control system by PPC Biofilter.

**biotower**    See "biological filter."

*Biotox*®    Regenerative thermal oxidation process by Biothermica International.

**bioturbation**    The net effect of the activity of benthic organisms at wastewater treatment plant discharges, which may aid in the dispersion of contaminants and increase the exchange of oxygen and nutrients between the sediment and water.

*Biox*™    Package water treatment plant by Bioscience, Inc.

*Bioxide*®    Odor control system by Davis Industries, Process Division.

**BIPM**    International Bureau of Weights and Measures.

*Birm*™    Granular filter media for removal of iron and manganese by Clack Corp.

*Bitumastic*®    Coal tar coating products by Carboline Company.

**bituminous coal**   A coal high in carbonaceous matter that yields a considerable amount of volatile waste matter when burned.

**black liquor**   Strong organic waste generated during kraft pulping process.

**black lung disease**   A lung disease caused by prolonged inhalation of coal dust, which results in fibrosis, or scarring of lung tissue.

**black sand**   Discoloration of filter sand resulting from manganese deposits.

*Blakeborough*   Former equipment manufacturer acquired by Brackett-Green, Ltd.

**blast furnace**   Furnace used in iron-making process in which hot blast air flows upward through the raw materials and exits at the furnace top.

*Blaw-Knox*   Former name of Buffalo Technologies, Inc.

**bleach**   Oxidizing compound, usually containing chlorine combined with calcium or sodium.

**bleed**   To draw accumulated liquid or gas from a line or container.

*Blendmaster*   Sludge mixer by McLanahan Corp.

*Blendrex*™   Motionless mixer by LCI Corp.

**blind flange**   A pipe flange with a blind end used to close the end of a pipeline.

**blinding**   The reduction or cessation of flow through a filter resulting from solids restricting the filter openings.

**BLM**   Bureau of Land Management.

**blood worm**   The larval stage of the midge fly.

**blowdown**   A discharge from a recirculating system designed to prevent a buildup of some material.

**blower**   Air conveying equipment that generates pressures up to 103 kPa (15 pounds per square inch), commonly used for wastewater aeration systems.

**blue baby syndrome**　See "methemoglobinemia."

**blue vitriol**　Common name for copper sulfate, used to control algae.

**BLWRS**　See "barriered landscape water renovation system."

**BMP**　See "best management practice."

**BNR**　Biological nutrient removal.

*Boa-Boom*　Oil spill containment booms by Environetics, Inc.

*Boat Clarifier®*　Boat-shaped wastewater treatment clarifier by United Industries, Inc.

**BOD**　See "biochemical oxygen demand."

**$BOD_5$**　Five-day carbonaceous or nitrification-inhibited biochemical oxygen demand. See also "biochemical oxygen demand."

**BODu**　See "ultimate BOD."

**body feed**　Coating material added to the influent of precoat filters during filtration cycle.

**bog**　Poorly drained land filled with decayed organic matter that is wet and spongy and unable to support any appreciable weight.

*BogenFilter™*　Belt filter press by Klein America, Inc.

**boiler**　A vessel in which water is continually vaporized into steam by the application of heat.

**boiler feedwater**　Water that, in the best practice, is softened and/or demineralized and heated to nearly boiler temperature and deaerated before being pumped into a steam boiler.

*Boilermate®*　Packed-column deaerator by Cleaver-Brooks.

**boiling point**　The temperature at which a liquid's vapor pressure equals the pressure acting on the liquid.

**boiling point elevation (BPE)**   The difference between the boiling point of a solution and the boiling point of pure water at the same pressure.

**boil out**   An evaporator cleaning process where wash water is boiled in an evaporator to remove scale deposits.

**bomb calorimeter**   An instrument used to determine the heat content of sludge or other material.

*BonoZon*   Ozone generator by ProMinent Fluid Controls, Inc.

**BOO**   Build, own, operate.

**boom**   A floating barrier used to contain oil on a body of water.

**booster pump**   A pump used to raise the pressure of the fluid on its discharge side.

**BOOT**   Build, own, operate, transfer.

*Boothwall*™   Cartridge filters for booth generated dust by Dustex Corp.

**bore hole**   A manmade hole in a geological formation.

*Bosker*   Trash rack cleaner by Brackett Green, Ltd.

**bottom ash**   The noncombustible particles that fall to the bottom of a boiler furnace.

**botulism**   A severe form of food poisoning usually associated with development of a toxic bacteria under anaerobic conditions in improperly preserved or prepared food.

**bound water**   Water held on the surface or interior of colloidal particles.

*Bowser-Briggs*   Former manufacturer of oil/water separation equipment.

**Boyle's law**   The volume of a gas varies inversely with its pressure at constant temperature.

*Boythorp*   Glass-coated steel tanks by Klargestor.

**BPE** See "boiling point elevation."

**BPR** (1) Biological phosphorus removal. (2) Boiling point rise.

**brackish water** Water containing low concentration of soluble salts, usually between 1000 and 10,000 mg/L.

**branch sewer** A sewer that receives wastewater from a small area and discharges into a main sewer serving more than one area.

*Brandol* Cylindrical fine-bubble diffusers by Schumacher Filters, Ltd.

**brass** A copper alloy containing up to 40% zinc.

**breakpoint chlorination** Addition of chlorine until the chlorine demand has been satisfied. Further addition will result in a chlorine residual so that disinfection can be assured.

**breakthrough** That point in the granular media filter cycle when the filtrate turbidity begins to increase because the filter bed is full and no longer able to retain solids.

**breakwater** An offshore barrier, often connected to shore, that breaks the force of waves and provides shelter from wave action.

*Breeze* Compact air stripping tank by Aeromix Systems, Inc.

**brine** Water saturated with, or containing a high concentration of, salts, usually in excess of 36,000 mg/L.

*Brinecell* Electrolytic generation unit by Brinecell, Inc.

**brine concentrator** A vertical tube falling film evaporator employing special scale control techniques to maximize concentration of dissolved solids.

**brine heater** The heat input section of a multistage flash evaporator where feedwater is heated to the process' top temperature.

**brine staging** See "reject staging."

**British thermal unit (Btu)**   The quantity of heat required to raise the temperature of 1 lb of water by 1° Fahrenheit.

**broad-crested weir**   A weir having a substantial crest width in the direction parallel to the direction of water flowing over it.

**broke**   Paper waste generated prior to completion of the papermaking process.

**bromine**   A halogen used as a water disinfectant in combination with chlorine as a chlorine-bromide mixture.

**bronze**   A copper-tin alloy, or any other copper alloy that does not contain zinc or nickel as the principal alloying element.

**brown coal**   A common term for lignite.

**Brownian motion**   Erratic movement of colloidal particles that results from the impact of molecules and ions dissolved in the solution.

*Brownie Buster*   Organic solids agitator/separator by Enviro-Care.

*Bruner-Matic*™   Treatment control center by Bruner Corp.

**brush aerator**   Mechanical aeration device most frequently used in oxidation ditch wastewater treatment plants, consisting of a horizontal shaft with protruding paddles that are rapidly rotated at the water surface. Also called a "rotor."

**BS&W**   Bottom sediments & water.

**BTEX**   Benzene, toluene, ethylbenzene, and xylene.

**Btu**   See "British thermal unit."

*BTU-Plus*®   Filter media that incinerates to inert ash by Alar Engineering Corp.

**BTX**   Benzene, toluene, and xylene.

**bubbler system**   Common terminology for pneumatic-type differential level controller.

**bubonic plague**   An acute infectious disease usually transmitted from infected animals to humans by the bite of a rat flea.

**buchner funnel**   A laboratory funnel with a perforated bottom that utilizes a disposable filter paper to evaluate wastewater and sludge dewaterability.

**bucket elevator**   A conveying device consisting of a head and foot assembly that supports and drives an endless chain or belt to which buckets are attached.

**buffer**   A substance that stabilizes the pH value of solutions.

**buffering capacity**   The capacity of a solution to resist a change in composition, especially changes in pH.

*Buflovak®*   Evaporator and crystallizer product line by Buffalo Technologies, Inc.

**bulk density**   The density/volume ratio for a solid including the voids contained in the bulk material.

**bulkhead**   A partition of wood, rock, concrete, or steel used for protection from water, or to segregate sections of tanks or vessels.

**bulking sludge**   A poorly settling activated sludge that results from the predominance of filamentous organisms.

**buoyancy**   The tendency of a body to rise or float in a liquid.

**BuRec**   U.S. Bureau of Reclamation.

**burette**   A glass tube with fine gradations and bottom stopcock used to accurately dispense fluids.

*Burger Press*   High-pressure belt filter press by EMO France.

**burning rate**   The rate at which solid waste is incinerated or heat is released during incineration.

**burnishing**   A surface finishing process in which surface irregularities are displaced rather than removed.

**bushing** (1) A short threaded tube that screws into a pipe fitting to reduce its size. (2) The bearing surface for pin rotation when a chain revolves around a sprocket.

**butterfly valve** A valve equipped with a stem-operated disk that is rotated parallel to the liquid flow when opened and perpendicular to the flow when closed.

*BVF®* Anaerobic wastewater treatment system by ADI Systems, Inc.

**BWI** British Drinking Water Inspectorate.

**bypass** A channel or pipe arranged to divert flow around a tank, treatment process, or control device.

**byproduct** A material or substance that is not a primary product of a process and is not separately produced.

# C

**CAA**   See "Clean Air Act."

**CAAA**   See "Clean Air Act Amendments."

**CableTorq**   Circular thickener with automatic torque load response system by Dorr-Oliver, Inc.

**CaCO₃**   See "calcium carbonate."

**CAD**   Computer-aided design.

*CADRE®*   Volatile organic compound destruction process by Vara International.

*Cairox®*   Potassium permanganate by Carus Chemical Co.

**caisson**   Watertight structure used for underwater work.

**cake**   Dewatered sludge with a solids concentration sufficient to allow handling as a solid material.

**cake filtration**   Filtration classification for filters where solids are removed on the entering face of the granular media.

*CakePress*   Modular high-pressure section of dewatering press by Parkson Corp.

**calandria**   The heating element in an evaporator consisting of vertical tubes that act as the heating surface.

**calcareous**   Composed of or containing calcium compounds, particularly calcium carbonate.

**calcify**   To become stone-like or chalky due to deposition of calcium salts.

**calciner**   A device in which the moisture and organic matter in phosphate rock is reduced in a combustion chamber.

**calcining**   Exposure of an inorganic compound to a high temperature to alter its form and drive off a substance that was originally part of the compound.

*Calciquest*   Liquid polyphosphate by Calciquest, Inc.

**calcium carbonate**   A white, chalky substance that is the principal hardness- and scale-causing compound in water. Chemical formula is $CaCO_3$.

**calcium carbonate equivalent (mg/L as $CaCO_3$)**   A convenient unit of exchange for expressing all ions in water by comparing them to calcium carbonate, which has a molecular weight of 100 and an equivalent weight of 50.

**calcium hypochlorite**   A chlorine compound frequently used as a water or wastewater disinfectant. Chemical formula is $Ca(OCl)$.

**calcium sulfate**   A white solid known as the mineral "anhydrite" with the chemical formula $CaSO_4$ and gypsum with the formula $CaSO_4 \cdot 2H_2O$.

*Caldicot Screen*   Self-cleaning bar screen by Advanced Wastewater Treatment Ltd.

**calorie**   The amount of heat required to raise the temperature of 1 gram of water at 15°C by 1°C.

*Calver*   Chemicals for use in the analysis of calcium in water by Hach Company.

*Calvert*   Manufacturer acquired by Monsanto Enviro-Chem Systems, Inc.

**CAMP**   Continuous air monitoring program.

*Camp Nozzle*   Plastic strainer-type nozzle for filter underdrain by Walker Process Equipment Co.

*Cannon™*   Positive displacement digester mixer by Infilco Degremont, Inc.

*Cannonball²*   Portable multiple gas detector by Biosystems, Inc.

*Cansorb*   Activated carbon adsorber by TIGG Corp.

**capillarity**   The ability of a soil to retain a film of water around soil particles and in pores through the action of surface tension.

**capillary action**   The action or movement of surface water, or water in very small interstices, due to the relative attraction of molecules of a liquid for each other and for those of a solid.

**capillary fringe**   The zone of porous material above the zone of saturation that may contain water due to capillarity.

*Capitox*   Modular wastewater treatment plant by Simon-Hartley, Ltd.

*Capozone*®   Ozone generation system by Capital Controls Co., Inc.

*Capsular*   Pump station by Smith & Loveless, Inc.

*Captivated Sludge Process*   Fixed-film biological waste treatment system by Waste Solutions.

*Captor*®   Fixed-film biological waste treatment system by Waste Solutions.

*CAR*™   Aerobic wastewater treatment system with covered reactor by ADI Systems, Inc.

*Carball*   Carbon dioxide generator formerly offered by Walker Process Equipment Co.

*Carbo Dur*™   Granular activated carbon by U.S. Filter Corp.

*Carbofilt*   Anthracite filter media by International Filter Media.

**carbon**   An element present in all materials of biological origin.

**carbon-14**   A naturally occurring radioactive isotope of carbon that emits beta particles when it undergoes radioactive decay.

**carbonaceous biochemical oxygen demand (CBOD)**   The portion of biochemical oxygen demand where oxygen consumption is due to oxidation of carbon, usually measured after a sample has been incubated for 5 days. Also called "first-stage BOD."

**carbon adsorption** The use of powdered or granular activated carbon to remove refractory and other organic matter from water.

**carbonate** A compound containing the anion radical of carbonic acid $CO_3$.

**carbonate alkalinity** Alkalinity caused by carbonate ions.

**carbonate hardness** The hardness in water caused by bicarbonates and the carbonates of calcium and magnesium.

**carbonation** The diffusion of carbon dioxide gas through a liquid.

**carbonator** A device used to carbonate or recarbonate water.

**carbon black** An additive that prevents degradation of thermoplastics by ultraviolet light.

**carbon cycle** A graphical presentation of the movement of carbon among living and nonliving matter.

**carbon dioxide** A noncombustible gas formed in animal respiration and the combustion and decomposition of organic matter. Chemical formula is $CO_2$.

**carbon fixation** A process occurring in photosynthesis where atmospheric carbon dioxide gas is combined with hydrogen obtained from water molecules.

*Carbonite* Anthracite filter media by Carbonite Filter Corp.

**carbon monoxide** A criterion air pollutant that is a colorless, odorless gas produced by incomplete combustion of organic fuels; it is lethal to humans at concentrations exceeding 5000 mg/L. Chemical formula is CO.

**carbon steel** A general purpose steel whose major properties depend on its 0.1 to 2% carbon content without substantial amounts of other alloying elements.

*Carborundum* Former manufacturer of wastewater treatment equipment.

**carboy** A large container used to store or transport liquid chemicals or water samples.

**carcinogen** A cancer- or tumor-causing agent.

*CA·RE*™ Spent cartridge filter recovery program by U.S. Filter Corp.

*Carela*™ Water treatment disinfectant products by R. Spane GmbH.

*Carrobic* Aerobic digester/thickener used with oxidation ditch wastewater treatment system by Eimco Process Equipment Co. (U.S.).

*Carrousel*® Biological oxidation/wastewater treatment system by DHV Water BV, licensed to Eimco Process Equipment Co. (U.S.).

*Cartermix* Anaerobic sludge digester mixing system by Ralph B. Carter Co.

**cartridge filter** Filter unit with cylindrical replaceable elements or cartridges.

*Carulite*® Catalysts for volatile organic compound destruction by Carus Chemical Co.

**Carver-Greenfield process** Multiple effect evaporation process to extract water from sludge.

*Cascade* Biological filtering system using synthetic media by Mass Transfer, Inc.

**casing** A pipe or tube placed in a bore hole to support the sides of the hole and to prevent other fluids from entering or leaving the hole.

**CaSO₄** See "calcium sulfate."

*CASS*™  Sequencing batch reactor wastewater treatment system by Transenviro, Inc.

**cast iron**  A general description for a group of iron-carbon-silicon metallic products obtained by reducing iron ore with carbon at temperatures high enough to render the metal fluid and cast it in a mold.

*CastKleen*  Cast-in-place filter underdrain by Eimco Process Equipment Co.

**catalyst**  A substance that modifies or increases the rate of a chemical reaction without being consumed in the process.

**catalytic converter**  A device installed in the exhaust system of an internal combustion engine that utilizes catalytic action to oxidize hydrocarbon and carbon monoxide emissions to carbon dioxide.

**catch basin**  An open basin that serves as a collection point for stormwater runoff.

**catchment**  A barrel, cistern, or other container used to catch water.

**catchment area**  The area of land bounded by watersheds draining into a river, lake, or reservoir.

**category I contaminant**  U.S. EPA contaminant category indicating that sufficient evidence of carcinogenicity via ingestion in humans or animals exists to warrant classification as "known or probable human carcinogens via ingestion."

**category II contaminant**  U.S. EPA contaminant category for which limited evidence of carcinogenicity via ingestion exists to warrant classification as "possible human carcinogens via ingestion."

**category III contaminant**  U.S. EPA contaminant category of substances for which insufficient or no evidence of carcinogenicity via ingestion exists.

**catenary bar screen**  Mechanical screening device using revolving chain-mounted rakes to clean a stationary bar rack.

*Cat Floc®*  Cationic polymer to enhance solids/liquid separation by Calgon Corp.

**cathode**  The negative electrode where the current leaves an electrolytic solution.

**cathodic protection**  Electrochemical corrosion protection achieved by imposing an electrical potential to counteract the galvanic potential between dissimilar metals which would lead to corrosion.

**cation**  A positively charged ion that migrates to the cathode when an electrical potential is applied to a solution.

**cationic polymer**  A polyelectrolyte with a net positive electrical charge.

*Cat-Ox™*  Catalytic oxidation system by Catalytic Combustion Corp.

**caustic**  Alkaline or basic.

**caustic soda**  Common term for sodium hydroxide. Chemical formula is NaOH.

**cavitation**  (1) A selective corrosion that results from the collapse of air or vapor bubbles with sufficient force to cause metal loss or pitting. (2) The action of a pump attempting to discharge more water than suction can provide.

**CBOD**  See "carbonaceous biochemical oxygen demand."

**CCB**  Coal combustion byproducts.

*CCC*  Streaming current coagulation control center by Milton Roy Co.

**CDC**  See "Centers for Disease Control."

*CDI™*  Continuous deionization process that regenerates resins with electricity by U.S. Filter Corp.

*CE-Bauer*   Former screening equipment supplier acquired by Andritz-Ruthner, Inc.

*Cecarbon*®   Granular activated carbon by Elf Atochem North America, Inc.

*Cecasorb*®   Adsorbent canisters containing activated carbon by Atochem North America, Inc.

*CELdek*   Synthetic media for evaporative cooling systems by Munters.

**CEM**   See "continuous emissions monitoring."

**CEMA**   Conveyor Equipment Manufacturers Association.

*CEMcat*™   Continuous emissions monitor by Advanced Sensor Devices, Inc.

**cement**   A powder that, mixed with water, binds a stone and sand mixture into strong concrete when dry.

**cementing**   The process of pumping a cement slurry into a drilled hole and/or forced behind the casing.

**cement kiln dust**   Alkaline material produced during the manufacture of cement that may be used to stabilize sludge.

**CEMS**   Continuous emissions monitoring system.

*Censys*™   Water and wastewater treatment products by U.S. Filter Corp.

*Centaur*™   Activated carbon by Calgon Carbon Corp.

**Centers for Disease Control (CDC)**   A U.S. Department of Health agency responsible for surveillance of disease patterns, developing disease control and prevention procedures, and public health education.

*Center-Slung*   Basket centrifuge by Ketema, Inc.

**centipose**   A unit of the dynamic viscosity of a liquid. The dynamic viscosity of water at 20°C is 1 centipose.

*Centrac*   Metering pump by Milton Roy Co.

*Centra-flo*™   Continuous-flow gravity sand filter by Applied Process Equipment.

**centrate** Dilute stream remaining in a centrifuge after solids are removed.

*Centric®* Overload clutch products by Zurn Industries, Inc.

*Centri-Cleaner®* Liquid cyclone by Andritz-Ruthner, Inc.

*Centridry™* Sludge dewatering and drying process by Humboldt Decanter, Inc.

*CentriField®* Wet scrubber by Entoleter, Inc.

**centrifugal pump** A pump with a high-speed impeller that relies on centrifugal force to throw incoming liquid to the periphery of the impeller housing where velocity is converted to head pressure.

**centrifugation** The use of centrifugal force to separate solids from liquids based on density differences.

**centrifuge** A dewatering device relying on centrifugal force to separate water and solids.

*Centripress™* Solid bowl centrifuge by Humboldt Decanter, Inc.

*CenTROL®* Gravity cluster sand filter by General Filter Co.

*Centrox®* Aspirating aerators by Hazelton Environmental Products, Inc.

*Cerabar* Pressure transmitter by Endress+Hauser.

*Ceraflo®* Ceramic membrane filters by U.S. Filter Corp.

**CERCLA** Comprehensive Environmental Response, Compensation and Liability Act. Also known as "Superfund."

**CERMS** Continuous emissions rate monitoring system.

**cesspool** A covered tank with open joints constructed in permeable soil to receive raw domestic wastewater and allow partially treated effluent to seep into the surrounding soil, while solids are contained and undergo digestion.

*CETCO*   Colloid Environmental Technologies Co.

**CFB**   Circulating fluidized bed.

**CFC**   See "chlorofluorocarbon."

**CFR**   Code of Federal Regulations.

**cfs**   Cubic feet per second.

**CFU**   See "colony forming units."

**CGMP**   Current good manufacturing practice.

*Chabelco*   Chain products marketed by Envirex, Inc.

**chain and flight collector**   A sludge collector mechanism utilized in rectangular sedimentation basins or clarifiers.

*Chainbelt*   Name of former Envirex, Inc., parent company.

*Chainsaver Rim*   Sludge collector sprockets with wear rim by Jeffrey Chain Co.

*Channelaire*™   Submersible aerator/mixer by Framco Environmental Technologies.

*Channel Flow*   Sewage disintegrator by C&H Waste Processing.

**channeling**   A condition that occurs in a filter or other packed bed when water finds furrows or channels through which it can flow without effective contact with the bed.

*Channel Monster*®   In-channel sewage grinder by JWC Environmental.

**characteristic hazardous waste**   A waste material declared hazardous because it exhibits ignitable, corrosive, reactive, or toxic characteristics.

**charge density**   In a polyelectrolyte, the mole ratio of the charged monomers to noncharged monomers.

**Charles' law**   The volume of gas at constant pressure varies in direct proportion to the absolute temperature.

**check valve**   A valve that opens in the direction of normal flow and closes with flow reversal.

*Check Well*   Well level measuring device by Drexelbrook Engineering Company.

**chelating agent**   A compound that is soluble in water and combines with metal ions to keep them in solution.

**chelation**   A chemical complexing of metallic cations with organic compounds to prevent precipitation of the metals.

*Chem-Flex®*   Portable holding tank by Aero Tec Laboratories, Inc.

*Chem-Gard®*   Direct and magnetically driven centrifugal pumps by Vanton Pump & Equipment Corp.

**chemical feeder**   A device used to dispense chemicals at a predetermined rate.

**chemical fixation**   The transformation of a chemical compound to a new, nontoxic form.

**chemical oxygen demand (COD)**   Measurement of organic matter in a water or wastewater that can be oxidized by using a chemical oxidizing agent.

**chemical sludge**   Sludge resulting from chemical treatment processes of inorganic wastes that are not biologically active.

*Cheminjector-D®*   Diaphragm pump series by Hydroflo Corp.

**chemisorption**   The formation of an irreversible chemical bond between the sorbate molecule and the surface of the adsorbent.

*Chemix*   Dry polymer mixing and feeding unit by Semblex, Inc.

*Chemomat*   Electrochemical membrane cell separation system by Ionics, Inc.

**chemotrophs**   Organisms that extract energy from organic and inorganic oxidation/reduction reactions.

*Chem-Scale*™   Weighing scale for vertical chemical tanks by Force Flow Equipment.

*ChemScan*   Process analyzers by Biotronics Technologies, Inc.

*ChemSensor*®   Volatile organic compound monitor by ORS Environmental Systems.

*Chemtact*™   Atomized mist odor scrubbing system by Quad Environmental Technologies Corp.

*Chem-Tower*®   Bulk chemical feeding unit by Smith & Loveless, Inc.

*Chemtrac*®   Streaming current monitoring systems by Chemtrac Systems, Inc.

*Chemtube*®   Diaphragm metering pump by Wallace & Tiernan, Inc.

*Chevron*™   Clarifier tube settlers by U.S. Filter Corp.

*Chicago Pump*   Product group of Yeomans Chicago Corp.

**chicane**   A plow or other obstacle used on a belt thickener or belt press to mix or turn sludge to facilitate sludge dewatering.

**chimney effect**   The tendency of air or gas in a vertical passage to rise when it is heated because its density is lower than the surrounding air or gas.

*Chi-X*®   Odor control product by NuTech Environmental Corp.

**chloramines**   Disinfecting compounds of organic or inorganic nitrogen and chlorine.

*Chlor-A-Vac*™   Gas induction systems by Capital Controls Co., Inc.

**chlorinated**   (1) The condition of water or wastewater that has been treated with chlorine. (2) A description of an organic compound to which chlorine atoms have been added.

**chlorination**   The addition of chlorine to a water or wastewater, usually for the purpose of disinfection.

**chlorinator**   A metering device used to add chlorine to water or wastewater.

**chlorine**   An oxidant commonly used as a disinfectant in water and wastewater treatment. Chemical formula is $Cl_2$.

**chlorine contact chamber**   A detention chamber to diffuse chlorine through water or wastewater while providing adequate contact time for disinfection.

**chlorine demand**   The difference in the amount of chlorine added to a water or wastewater and the amount of residual chlorine remaining after a specific contact duration, usually 15 min.

**chlorine residual**   The amount of chlorine remaining in water after application at some prior time. See "free chlorine residual."

**chlorine tablets**   Common term for pellets of solidified chlorine compounds such as calcium hypochlorite used for water disinfection.

*ChlorMaster*™   Sodium hypochlorite generation system by Pepcon Systems, Inc.

**chlorofluorocarbon (CFC)**   Any of the ozone-depleting compounds containing carbon and one or more halogens, usually fluorine, chlorine, or bromine, which have been used as commercial refrigerants and propellants in aerosol sprays.

**chloroform**   A trihalomethane formed by the reaction of chlorine and organic material in water. Chemical formula is $CHCl_3$.

*Chloromatic*   Electrolytic chlorine generator by Brinecell, Inc.

*Chloropac*®   Hypochlorite generation system by Electro-catalytic, Ltd. (Europe) and ELCAT Corporation (U.S.).

*Chlor-Scale™*   Weighing device for chlorine ton containers by Force Flow Equipment.

*Chlortrol*   Residual chlorine analyzer by Fisher & Porter.

**cholera**   Highly infectious disease of the gastrointestinal tract caused by waterborne bacteria.

*Chopper Pump*   Pump that chops solids between the impeller and fixed cutter bar by Vaughn Company, Inc.

**C-horizon**   The unaltered soil layer underlying the B-horizon containing a minimum of soil fauna and flora.

**chromatography**   The separation of a mixture into its component compounds according to their relative affinity for a solvent system or column media.

*Chromaver*   Chemical reagents used to determine presence of chromates in solutions by Hach Company.

**chronic toxicity test**   Test method used to determine the concentration of a substance that produces an adverse effect on a test organism over an extended period of time.

**Ci**   See "curie."

*Cide-Trak™*   Biocide monitoring system by Microbics Corp.

*CirculAire™*   Aspirating aerator by Aeration Industries, Inc.

*Circuline*   Circular sludge collector product line by FMC Corp., MHS Division.

*Circumfed*   Dissolved air flotation unit by Tenco Hydro, Inc.

**cistern**   A small covered tank for storing water, usually placed underground.

**citric acid**   A crystalline acid present in citrus fruits. Chemical formula is $C_6H_8O_7 \cdot H_2O$.

**CIWEM**   The Chartered Institution of Water and Environmental Management.

*CIX*™ Ion exchange system by Kinetico Engineered Systems, Inc.

Cl₂ See "chlorine."

*CLAM*® Cleansimatic liquid analysis meter by Monitek Technologies, Inc.

*Claraetor* Circular clarifier with aeration compartment formerly offered by Dorr-Oliver, Inc.

*ClarAtor* Clarifier technology by Aero-Mod, Inc.

*ClariCone*™ Solids contact clarifier by CBI Walker, Inc.

clarifier A quiescent tank used to remove suspended solids by gravity settling. Also called sedimentation or settling basins, they are usually equipped with a motor-driven rake mechanism to collect settled sludge and move it to a central discharge point.

*Clari-Float*® Package wastewater treatment plant including dissolved air flotation by Tenco Hydro, Inc.

*Clarifloc*® Polyelectrolyte used to enhance liquid/solid separation by Polypure, Inc.

*Clariflocculator* Combination clarifier and flocculator by Dorr-Oliver, Inc.

*ClariFlow* Upflow clarifier products by Walker Process Equipment Co.

*Clarigester* Two-story tank combining clarification and digestion by Dorr-Oliver, Inc.

*Clarion* Absorption media by Colloid Environmental Technologies Co.

*Clar+Ion*® Cationic coagulants and flocculants by General Chemical Corp.

*Claripak* Upflow, inclined plate clarifier by Aerators, Inc.

*ClariShear*™ Floating sludge collector by Techniflo Systems.

*ClariThickener*™ Combination clarifier and thickener by Eimco Process Equipment Co.

*Clari-Trac®*   Track-mounted siphon sludge removal system for rectangular clarifiers by F.B. Leopold Co., Inc.

*Clari-Vac®*   Floating bridge type siphon sludge removal unit or rectangular clarifiers by F.B. Leopold Co., Inc.

*Clar-i-vator®*   Solids contact clarifier by Smith & Loveless, Inc.

*Clar-Vac*   Induced and dissolved air flotation systems by Dontech, Inc.

**classifier**   A device used to separate constituents according to relative sizes or densities.

**clathrate**   A compound formed by the inclusion of molecules in cavities formed by crystal lattices.

**clay**   A fine-grained earthy material that is plastic when wet, rigid when dried, and vitrified when fired to high temperatures.

**clay liner**   A layer of clay soil added to the bottom and sides of an earthen basin for use as a disposal site of potentially hazardous wastes.

**Clean Air Act (CAA)**   1970 U.S. federal law requiring air pollutant emission standards; reauthorized in 1977 and 1990.

**Clean Air Act Amendments (CAAA)**   Amendments issued in 1990 to expand the EPA's enforcement powers and place restrictions on air emissions.

*Clean-A-Matic*   Self-cleaning basket strainer by GA Industries, Inc.

*Clean Chemicals*   High-purity chemicals for laboratory use only by Hach Company.

*Clean Shot*   Pneumatic solids delivery system by Wheelabrator Engineered Systems, Inc., CPC Engineering Products.

**Clean Water Act (CWA)**   1972 U.S. federal law regulating surface water discharges; updated in 1987.

*Clearcon* Circular clarifier product line by Vulcan Industries, Inc.

**clear cutting** The practice of completely felling a stand of trees, usually followed by the replanting of a single species.

*Clearflo* Cylindrical clarifier by Roberts Filter Manufacturing Co.

*Clear View*™ Continuous emissions monitoring system by Goal Line Environmental Technologies.

**clearwell** A tank or reservoir of filtered water used to backwash a filter.

*Climber*® Reciprocating rake bar screen by Infilco Degremont, Inc. (U.S.) and Hellmut Geiger GmbH (Germany).

*ClimbeRack*™ Bar screen gear rack that eliminates need for lubrication by Infilco Degremont, Inc.

**clinker** A fused byproduct of the combustion of coal or other solid fuels.

**clino** See "clinoptilolite."

**clinoptilolite** A naturally occurring clay that can be used in an ion exchange process for ammonia removal.

*Cloromat* Package chlorine/caustic plant by Ionics, Inc.

**close-coupled pump** A pump coupled directly to a motor without gearing or belting.

**closed cycle cooling system** A cooling water system in which heat is transferred by recirculating water contained within the system, producing a relatively small blowdown stream of concentrated solids.

*Clostridium botulinum* Anaerobic microbe that causes botulism.

**closure plan** Written plan to decommission and secure a hazardous waste management facility.

**cloud** A mass of small water droplets in the atmosphere that are not of sufficient size to fall to the earth.

**cloud seeding** The artificial introduction of chemicals such as silver iodide or dry ice into clouds to induce rain.

*CLR Process* Closed-loop reactor oxidation ditch process by Lakeside Equipment Corp.

**CMA** Chemical Manufacturers Association.

**$CO_2$** See "carbon dioxide."

*Coagblender* Turbine-type in-line and open-channel mixers by Aerators, Inc.

**coagulation** The destabilization and initial aggregation of finely divided suspended solids by the addition of a polyelectrolyte or a biological process.

**coalesce** The merging of two droplets to form a single, larger droplet.

**coal gasification** The conversion of solid coal to a gas mixture to be used as fuel.

**coal pile runoff** Rainfall runoff from or through a coal storage pile.

**Coanda effect** The tendency of a liquid coming out of a nozzle or orifice to travel close to the wall contour even if the wall curves away from the jet's axis.

**coarse-bubble aeration** An aeration system that utilizes submerged diffusers which release relatively large bubbles.

**coarse sand** Sand particles, usually larger than 0.5 mm.

**coarse screen** A screening device usually having openings greater than 25 mm (1″).

**coastal reclamation** Reclaiming land from shallow coastal areas of the sea by dumping rubble and refuse, or by constructing breakwaters and sea walls and drainage of the enclosed area.

**Coastal Zone Management Act (CZMA)** Act requiring all federal agencies and permittees who conduct activities affecting a state's coastal zone to comply with an approved state coastal zone management program.

**COC**   See "cycles of concentration."

**cocci**   Sphere-shaped bacteria.

**COD**   Chemical oxygen demand.

*CodeLine*™   Membrane pressure vessel housing by Advanced Structures, Inc.

**codisposal**   A method of sludge disposal where the digested sludge is mixed with sorted refuse and incinerated, composted, or treated by pyrolysis prior to final disposal.

**coefficient of haze (COH)**   A measure of air visibility determined by the darkness of the stain remaining on white paper after it has been used to filter air.

*Coex Seal*™   Containment liner by National Seal Co.

**cofferdam**   A temporary dam, usually of sheet piling, built to provide access to an area that is normally submerged.

*CogBridge*   Traveling bridge sludge collector by Walker Process Equipment Co.

**cogen**   See "cogeneration."

**cogeneration**   A power system that simultaneously produces both electrical and thermal energy from the same source.

*Cog Rake*   Reciprocating rake bar screen by FMC Corp., MHS Division.

**COH**   See "coefficient of haze."

**cohort study**   An epidemiological study where population subgroups with a common exposure to a suspected disease-causing agent are studied over time to determine the risk of developing disease.

*Coilfilter*   Rotary vacuum belt filter by Komline-Sanderson Engineering Corp.

**coke**   The solid carbon residue resulting from the distillation of coal or petroleum.

**coke tray aerator**   An aerator where water is sprayed or flows over coke-filled trays.

**cold lime-soda softening**  Lime-soda softening process of water treatment at ambient temperatures.

*Cold Plasma*®  Ozone disinfection systems by Cimco Lewis Ozone Systems, Inc.

**coliform bacteria**  Rod-shaped bacteria living in the intestines of humans and other warm-blooded animals.

*Colilert*®  Reagent used to detect and identify coliforms and *E. coli* by IDEXX Laboratories, Inc.

*ColiSure*  Coliform presence/absence test medium by Millipore.

*Collectaire*  Airlift activated sludge removal system by FMC Corp., MHS Division.

**collection main**  The public sewer to which a building service or individual system is connected.

**collector chain**  Chain used to convey sludge scraper in a rectangular sludge collector.

*Collision Scrubber*™  Air pollution control scrubber by Monsanto Enviro-Chem Systems, Inc.

**colloid**  Suspended solid with a diameter less than 1 micron that cannot be removed by sedimentation alone.

*Colloidair Separator*™  Open-basin dissolved air flotation by U.S. Filter Corp.

**colony forming units (CFU)**  The number of bacteria present in a sample as determined in a laboratory plate count test where the number of visible bacteria colony units present are counted.

**color**  Water condition resulting from presence of colloidal material (see "apparent color") or organic matter (see "true color"), measured by visual comparison with lab prepared standards.

**colorimeter**  A photoelectric instrument used to measure the amount of light of a specific wavelength absorbed by a solution.

*Color-Katch*™  Flocculant/coagulant by Kem-Tron.

*ColOX*™ Fixed-film aerobic bioreactor system by TETRA Technologies, Inc.

*Combi-Guard* Packaged screening unit by Andritz Sprout-Bauer S.A.

**combination chain** Chain used in conveyor applications, having cast block links with steel pins and connecting bars.

**combined available chlorine** The concentration of chlorine combined with ammonia as chloramine and still available to oxidize organic matter.

**combined cycle generation** A gas turbine generator system where heat from turbine generator exhaust gases are recovered by a steam generating unit whose steam is used to drive a steam turbine generator.

**combined sewer** A sewer used to receive sanitary wastewater, stormwater, and surface water.

**combined sewer overflow (CSO)** Wastewater flow that consists of stormwater and sanitary sewage.

*Combu-Changer*® Regenerative oxidizer to control volatile organic compound emissions by ABB Air Preheater, Inc.

**combustibles** Materials that can be ignited at a specific temperature in the presence of air to release heat.

**combustion gases** The mixture of gases and vapors produced by burning.

*Combustrol*® Fly ash conditioning treatment technology marketed by Wheelabrator Clean Air Systems, Inc.

*Comet* Electrically driven rotary distributor for a fixed-film reactor by Simon-Hartley, Ltd.

**commercial use** Use in commercial enterprise providing salable goods or services.

**commercial waste** Solid waste from nonmanufacturing establishments such as office buildings, markets, restaurants, and stores.

**comminutor**    A circular screen with cutters that grinds large sewage solids into smaller, settleable particles.

**community water system**    A public water system serving at least 25 year-round residents or that has 15 or more connections used by year-round residents.

*Compact CDI*™    Continuous deionization product by U.S. Filter Corp.

*Compact RO*    Reverse osmosis product by U.S. Filter Corp.

*Completaire*    Package waste treatment plant with complete mix activated sludge by FMC Corp.

*CompleTreator*    Package trickling filter waste treatment plant formerly offered by Dorr-Oliver, Inc.

*Compmaster*™    Composting system computer process control system by Wheelabrator Clean Water Systems, Inc., Bio Gro Division.

*Component Clarifier*    Standard group of clarifier components and options that can be matched with an application's requirements by Eimco Process Equipment Co.

**composite sample**    A water or wastewater sample made up of a number of samples taken at regular intervals over a 24-hour period.

**compost**    The endproduct of composting.

*Compost-A-Matic*    Sludge composting system by Farmer Automatic of America, Inc.

**composting**    Sludge stabilization process relying on the aerobic decomposition of organic matter in sludge by bacteria and fungi.

*Compost Storm Water Filter*    Stormwater filtration/treatment system by CSF Treatment Systems, Inc.

**compound**    A substance consisting of two or more independent elements that can be separated only by a chemical reaction.

*Compound 146*    Polyurethane material used for sprocket tooth inserts by FMC Co.

**compression settling** Phenomenon referring to sedimentation of particles in a concentrated suspension where further settling can occur only by compression of the existing structure of settled particles.

**compressor** A mechanical device used to increase the pressure of a gas or vapor.

**concentration** (1) The amount of a substance dissolved or suspended in a unit volume of solution. (2) The process of increasing the amount of a substance per unit volume of solution.

**concentration polarization** A phenomenon in which solutes form a dense, polarized layer next to a membrane surface, which eventually restricts flow through the membrane.

**concentration ratio** The ratio of the concentration of solids in a water system to those of the dilute makeup water added to the system.

**concrete** A mixture of water, sand, stone, and a binder that hardens to a stone-like mass.

**condensate** Water obtained by evaporation and subsequent condensation.

**condensate polishing** Treatment of condensate water to achieve required purity.

**condensation** The change in state from vapor to liquid; the opposite of evaporation.

*Condense-A-Hood* Air/odor collecting hoods by Bedminster Bioconversion Corp.

**condenser** A heat transfer device used to cool steam and convert it from the vapor to liquid phase.

**conditioning** Pretreatment of a wastewater or sludge, usually be means of chemicals, to facilitate removal of water in a subsequent thickening or dewatering process.

**conductance** A measure of a solution's electrical conductivity that is equal to the reciprocal of the resistance.

**conduction**   The transfer of heat from one body to another by direct contact.

**conductivity**   The ability of a substance to conduct electricity; directly related to the mineral content of water.

*Cone Screen*   Internally fed rotary fine screen by Andritz-Ruthner, Inc. (Western Hemisphere) and Contra-Shear Engineering, Ltd.

*Configurator®*   Computerized process design and equipment selection tool by U.S. Filter Corp.

**confluence**   The point where the flow of streams or rivers meet.

**congeal**   To thicken, jell, or solidify, usually by cooling or freezing.

**connate water**   Water trapped in sedimentary rocks during their formation. Also known as "fossil water."

*Conoscreen*   Rotating disc microscreen by Purator Waagner-Brio.

**consent decree**   A binding agreement by two parties in a lawsuit that settles all questions raised in the case and does not require additional judicial action.

**constant-rate filtration**   Filter operation where flow through the filter is maintained at a constant rate by an adjustable effluent control valve.

**constructed wetlands**   A wastewater treatment system using the aquatic root system of cattails, reeds, and similar plants to treat wastewater applied either above or below the soil surface.

**consumptive waste**   Water that returns to the atmosphere without beneficial use.

*Contaclarifier*   Upflow buoyant media clarifier by Roberts Filter Co.

*Contac-Pac*   Circular steel contact aeration package waste treatment plant by FMC Corp.

**contact condenser** A device in which steam is condensed through direct contact with a cooling liquid.

**contact process** Wastewater treatment process where diffused air is bubbled over fixed media surfaces.

**contact stabilization process** Modification of the activated sludge process where raw wastewater is aerated with activated sludge for a short time prior to solids removal and continued aeration in a stabilization tank. Also called "biosorption process."

**contaminant** Any foreign component present in another substance.

**contamination** The degradation of natural water, air, or soil quality resulting from human activity.

*Cont-Flo*™ Back-cleaned reciprocating rake bar screen by John Meunier, Inc.

*Continental*® Water treatment products and systems by U.S. Filter Corp.

**contingency plan** A document setting out an organized, planned, and coordinated course of action to be followed in case of fire, explosion, or release of hazardous waste constituents, which could threaten human health or the environment.

**contingent valuation survey** (CVM) A survey technique for assigning value to injured natural resources based on respondents' willingness to support various resources in monetary terms.

**continuous emissions monitoring** The continuous measurement of pollutants emitted into the atmosphere from combustion or industrial processes.

*Continuous-Flo* Traveling bridge filter by Zimpro Environmental, Inc.

**contract operations** Private operation of municipal facilities such as water and wastewater treatment plants.

*Contraflo*® Solids contact clarifier by General Filter Co.

*Contraflux*®   Countercurrent activated carbon adsorption unit by Graver Co.

*Contra-Shear*®   Screening equipment product line by Andritz Ruthner, Inc. (Western Hemisphere) and Contra-Shear Engineering, Ltd.

*Contreat*®   Aerobic wastewater treatment package plant by EnviroSystems Supply, Inc.

**convection**   The transfer of heat by a moving fluid such as air or water.

*Convertofuser*®   Wide-band coarse-bubble diffuser with fine sheath by FMC Corp.

**cooling pond**   A pond where water is cooled by contact with air prior to reuse or discharge.

**cooling tower**   An open water recirculating device that uses fans or natural draft to draw or force ambient air through the device to cool warm water by direct contact.

**cooling tower blowdown**   A sidestream of water discharged from a cooling tower recirculation system to prevent scaling or precipitation of saturated salts or minerals.

**cooling water**   Water used, usually in a condenser, to reduce the temperature of liquids or gases.

*COP*   Clarifier optimization program by WesTech Engineering, Inc.

*CopaClarifier*   Secondary clarifiers with filter brushes by Hydro-Aerobics (U.S.) and Copa Group (U.K.).

*Copa-NILL*   Tipping bucket tank flush system by Hydro-Aerobics (U.S.) and Copa Group (U.K.).

*Copasacs*   Fine screening sack by Hydro-Aerobics (U.S.) and Copa Group (U.K.).

*Copa Screen*   Packaged screening and dewatering unit by Longwood Engineering Co.

*Copasocks*   Sock-type screening sack by Hydro-Aerobics (U.S.) and Copa Group (U.K.).

*Copatrawl*   Sock-type screening sack by Hydro-Aerobics (U.S.) and Copa Group (U.K.).

*Copawash*   Rotating boom wash system for stormwater tanks by Hydro-Aerobics (U.S.) and Copa Group (U.K.).

*CopaWets*   Chemical coagulation and flocculation process for wastewater treatment by Copa Group.

**coping**   The top or covering of an exterior masonry wall.

*Coplastix®*   Synthetic composite used in the fabrication of sluice gates and stop logs by Ashbrook Corp. (U.S.) and Simon-Hartley, Ltd. (U.K.).

**copperas**   Common name for ferrous sulfate heptahydrate, a common coagulant. Chemical formula is $FeSO_4 \cdot 7H_2O$.

**copper-nickel**   A copper alloy containing 10–30% nickel to increase resistance to corrosion and stress corrosion cracking. Also called "cupronickel."

**copper sulfate**   Chemical used for algae control, also called "blue vitriol." Chemical formula is $CuSO_4$.

**corner sweep**   Scraper used to remove sludge from the corner of a square clarifier.

*Corosex®*   Processed magnesia used in filters to neutralize acidity by Clack Corp.

*Cor-Pak®*   Catalytic oxidizer system by ABB Air Preheater, Inc.

**corrosion**   Attack on material through chemical or electrochemical reaction with surrounding medium.

**corrosive**   The characteristic of a chemical agent that reacts with the surface of a material causing it to deteriorate or wear away.

**corrugated plate interceptor (CPI)**   Oil separation device utilizing inclined corrugated plates to separate free nonemulsified oil and water based on their density difference.

*Corten*   High-strength, low-alloy steel with enhanced atmospheric corrosion resistance by U.S. Steel Corp.

*Counter Current®*   Aeration process using rotating diffusers suspended from a rotating bridge by Schreiber Corp.

**coupon test**   A method of determining the rate of corrosion or scale formation by placing metal strips, or coupons, of a known weight in a tank or pipe.

**cover material**   Soil or other suitable material used to cover solid wastes in a sanitary or secure landfill.

*Covertite*   Clear span wastewater treatment tank cover by Thermacon Enviro Systems, Inc.

*C. parvum*   See "Cryptosporidium."

*CPC*   Conveying products group of Wheelabrator Engineered Systems, Inc., CPC Engineering.

CPI   Chemical process industry.

CPI   See "corrugated plate interceptor."

CPM   Critical path method.

CPVC   Chlorinated polyvinyl chloride. A chlorinated form of PVC that provides increased heat resistance.

**cradle-to-grave**   Hazardous waste management concept that attempts to track hazardous waste from its generation point (cradle) to its ultimate disposal point (grave).

*Crane®*   Water treatment product line by Cochrane Environmental Systems.

**crenothrix**   See "iron bacteria."

**crevice corrosion**   Localized corrosion in narrow crevices filled with liquid.

**criteria pollutants**   The major air pollutants, including carbon monoxide, hydrocarbons, lead, nitrogen dioxide, sulfur dioxide, ozone, and suspended particulates, for which the U.S. EPA has established ambient air quality standards.

**critical flow** The rate of flow of a fluid equal to the speed of sound in that fluid.

**critical pitting temperature** A value used to compare a material's resistance to pitting corrosion.

**critical point** The combination of pressure and temperature at which point a gas and liquid become indistinguishable.

*Cromaglass* Batch treatment wastewater system by Cromaglass Corp.

**cross-collector** A mechanical sludge collector mechanism, extending the width of one or more longitudinal sedimentation basins, used to consolidate and convey accumulated sludge to a final removal point.

**cross-connection** A physical connection in a plumbing system through which a potable water supply could be contaminated.

*Cross/Counteflo* Inclined-plate clarifier by Zimpro Environmental, Inc.

*Cross-Flo* Inclined static screen by Kason Corp.

**crossflow filtration** Method of filtration where the feedwater flows parallel to the surface of the filter medium.

*Crossflow Fouling Index*™ Membrane fouling test index by Argo Scientific.

**crosslinkage** The degree of bonding of a monomer or set of monomers to form an insoluble, three dimensional resin matrix.

*Crouzat*™ Water treatment products and systems by U.S. Filter Corp.

*Crown*™ Self-priming sewage pump product line by Crane Pumps & Systems.

*Crown Press*™ Sludge dewatering test device by Neogen Corp.

*CRP*® Continuous recirculation sludge mixing process for anaerobic digesters by FMC Corp., MHS Division.

**CRT**    Cell residence time.

**crude oil**    Unrefined petoleum as produced from underground formations.

**crumb rubber**    Ground or shredded rubber produced by shredding used automobile tires; it can be recycled in asphalt-rubber or other products.

**crypto**    See "Cryptosporidium."

**cryptosporidiosis**    Gastrointestinal disease caused by the ingestion of waterborne *Cryptosporidium parvum*, often resulting from drinking water contaminated by runoff from pastures or farmland.

**Cryptosporidium**    A protozoan parasite that can live in the intestines of humans and animals.

*Cryptosporidium parvum*    A species of Cryptosporidium known to be infective to humans.

**crystal**    A homogenous chemical substance that has a definite geometric shape, with fixed angles between its faces and distinct edges or faces.

**crystalline**    Having a regular molecular structure evidenced by crystals.

**crystallization**    The process of forming crystals.

**crystallizer**    Common term for a forced circulation evaporator.

**CSA**    Canadian Standards Association.

*CSF*™    Compost stormwater filter by CSF Treatment Systems, Inc.

**CSO**    Combined sewer overflow.

**CSTR**    Completely stirred tank reactor.

**cubic meter (m³)**    A volume measurement equal to 1000 L or 264. 2 gallons. One cubic meter of water weighs approximately 1 metric ton.

*Cullar*®    Activated carbon filter by Culligan International Corp.

*Cullex*    Softening resin by Culligan International Corp.

***Cullsorb***   Greensand filter by Culligan International Corp.

**culm**   Coal dust or anthracite tailings.

**culture**   A microbial growth developed by furnishing sufficient nutrients in a suitable environment.

**culvert**   An enclosed channel serving as a continuation of an open stream where a stream meets a roadway or other barrier.

**cupric**   Of or containing copper.

**cupric sulfate**   Copper sulfate.

**cupronickel**   See "copper-nickel."

**cup screen**   A single-entry, double-exit drum screen.

**curie (Ci)**   A unit of radioactivity equal to $3.7 \times 10^{10}$ disintegrations per second.

***Currie Clarifier***   Circular clarifier with aeration compartment formerly offered by Dorr-Oliver, Inc.

**curtain wall**   An external wall that is not load bearing. Usually refers to a wall that extends down below the surface of the water to prevent floating objects from entering a screen forebay.

***Curveco***   Mechanically cleaned bar screen by Eco Equipment.

***Cutrine-Plus***   Algicide/herbicide by Applied Biochemists, Inc.

***Cuver***   Chemical used to detect waterborne copper by Hach Company.

**CVM**   See "contingent valuation survey."

**CWA**   See "Clean Water Act."

**cwt**   Hundredweight.

***Cyanamer®***   Scale inhibitors and dispersants by Cytec Industries, Inc.

**cyanazine**   A common, and potentially carcinogenic, herbicide sometimes found in drinking water.

**cyanide**  A compound containing a CN group, usually extremely poisonous, often used in electroplating and other chemical processes.

*Cybreak*™  Emulsion breaker by American Cyanamid Co.

*Cycle-Let*®  Wastewater treatment and recycling system by Thetford Systems, Inc.

**cycles of concentration (COC)**  The ratio of the total dissolved solids concentration in a recirculating water system to the total dissolved solids concentration of the makeup water.

*Cyclesorb*®  Granular activated carbon adsorption system by Calgon Carbon Corp.

*Cyclo Blower*  Air blower by Gardner-Denver, Division of Cooper Industries.

*CycloClean*™  Hydrocyclone separator by Krebs Engineers.

*Cyclofloc*  Method for increasing clarifier rise rates by Biwater-OTV Ltd.

*Cy-Clo-Grit*  Prefabricated cyclonic grit collector by Jones and Attwood, Inc.

*Cyclo Grit Washer*  Inclined screw-type grit washer and dewatering unit by Eimco Process Equipment Co.

*Cyclo-Hearth*  Multiple hearth furnace by Zimpro Environmental, Inc.

*Cyclone*  Coarse-bubble diffuser by Aeromix Systems, Inc.

*Cyclo/Phram*®  Rotary plunger metering pump by Leeds & Northrup.

*Cyclotherm*  Sludge heat exchanger by FMC Corp., MHS Division.

*Cyclo-Treat*™  Cyclone separator by Envirex, Inc.

*Cygnet*  Rotary distributor for a fixed-film reactor by Simon-Hartley, Ltd.

**cyst**    A resting stage formed by some bacteria and protozoa in which the whole cell is surrounded by a protective layer.

**cytotoxin**    Any material toxic to cells.

**CZMA**    See "Coastal Zone Management Act."

# D

**D-20**  Filter underdrain nozzle by Infilco Degremont, Inc.

*Dac Floc*  Polyelectrolyte used to enhance liquid/solid separation by Dacar Chemical Co.

**DAF**  See "dissolved air flotation."

**DAFT**  Dissolved air flotation thickener.

**daily cover**  Cover material spread and compacted on the top and side slopes of compacted solid waste at the end of each day to control fire, moisture, and erosion, and to ensure an aesthetic appearance.

*Dakota*  Belt filter press by HydroCal Company.

**DALR**  Dry adiabatic lapse rate. See "adiabatic lapse rate."

**dalton**  A nominal unit of weight equal to that of a single hydrogen atom; $1 \times 10^{-24}$ grams.

**Dalton's law of partial pressure**  In a mixture of gases, each gas exerts pressure independently of the others, and the pressure of each gas is proportional to the amount of that gas in the mixture.

*Dano Drum*™  Rotating composting vessel by Reidel Smith Environmental, Inc.

**darcy**  A unit of measure used to indicate permeability, standardized by the American Petroleum Institute.

*DataGator*™  Sewer flow metering system by TN Technologies, Inc.

*DataRAM*™  Continuous readout monitor of airborne particles by MIE, Inc.

*DAVCO*  Municipal waste systems division of Davis Water & Waste Industries, Inc.

*Davy Bamag*  Former name of Bamag GmbH.

**day tank** Tank used to store chemicals or diluted polymer solution for 24 hr or less.

**dB** See "decibel."

*DBC Plus®* Bacterial culture for wastewater treatment by Enviroflow Inc.

**DBP** See "disinfection byproducts."

*DBS™* Clarifier and thickener drive units by DBS Manufacturing, Inc.

**DC** Direct current.

*D-Chlor™* Dechlorination system using sodium sulfite tablets by Eltech International Corp.

*DCI System Six™* Open-channel UV system by Fischer & Porter.

**DCS** See "distributed control system."

**D/DBP** Disinfectants, disinfection byproducts.

**D/DBP Rule** Proposed U.S. EPA rule to limit the maximum contaminant level of trihalomethanes.

**DDT** A chlorinated hydrocarbon insecticide, banned in many countries because of its persistence in the environment and accumulation in the food chain.

**DE** See "diatomaceous earth."

**deaerator** Device used to remove dissolved gases from solution.

**dealkalization** Any process that removes or reduces alkalinity of water.

**dealkalizer** Ion exchange unit with strong anion bed used to reduce bicarbonate alkalinity.

*DeAmine™* Odor control product by NuTech Environmental Corp.

**deashing** See "demineralizing."

**decant** Separation of a liquid from settled solids by pouring or drawing off the upper layer of liquid after the solids have settled.

**decarbonator**   A device used to remove alkalinity from solution by conversion to $CO_2$ prior to air stripping.

*DeCelerating Flo*   Gravity sand filter by CBI Walker, Inc.

**dechlorination**   The partial or complete reduction of chlorine by a physical or chemical process.

**decibel (dB)**   The unit for measuring sound pressure level.

**declining-rate filtration**   Filter operation where the rate of flow through the filter declines and the level of the liquid above the filter bed rises throughout the length of the filter run.

**decontamination**   The process of reducing or eliminating the presence of harmful substances, such as infectious agents, so as to reduce the likelihood of disease transmission from those substances.

*Dee Fo™*   Foam control additive by Ultra Additives, Inc.

*DeepAer*   Aeration system with rectangular eductor tubes by Walker Process Equipment Co.

*DeepBed*   Granular media filter bed used as fixed-film reactor by TETRA Technologies, Inc.

**deep bed filter**   Granular media filter with a sand or anthracite filter bed up to 1.8 m (6 ft) deep.

*Deep Bubble™*   Corrosion control aeration system by Lowry Aeration Systems, Inc.

*Deep Draw*   Airlift diffuser for lagoons by Wilfley Weber, Inc.

**deep well injection**   Disposal technique where raw or treated wastes are discharged through a properly designed well into a geological stratum.

*Defined Substrate Technology™*   Reagent system designed to promote growth of target microbe by Environetics, Inc.

*Deflectofuser®*   Coarse-bubble air diffuser by FMC Corp., MHS Division.

**defoliant**   A chemical applied to plants that causes them to lose their leaves.

**deforestation**   The permanent clearing of forest land and its conversion to nonforest uses.

**degasifier**   Device used to remove dissolved gases from solution, usually by means of an air stripping column.

**degradation**   The biological breakdown of organic substances.

**dehydrate**   The physical or chemical process where water in combination with other matter is removed.

*DeHydro*®   Vacuum-assisted sludge drying bed by Infilco Degremont, Inc.

**deinking**   The process of removing ink from secondary fibers.

**deionization (DI)**   The process of removing ions from water, most commonly through an ion exchange process.

*dekSPRAY*   Cooling tower nozzles by Munters.

*Delaval Filter*   Precoat condensate polishing filter by Idreco USA, Ltd.

**deliquescent**   The ability of a dry solid to absorb water from the air.

*DelPAC*   Polyaluminum coagulants by Delta Chemical Corp.

*Delrin*   High-molecular-weight acetal resin polymer material by E.I. DuPont De Nemours & Co.

**delta**   The flat alluvial area at the mouth of some rivers where an accumulation of river sediment is deposited in the sea or a lake.

*Delta-G*®   Parallel-plate separator by Smith & Loveless, Inc.

**delta P**   Differential pressure.

*Deltapilot*   Hydrostatic level measurement device by Endress+Hauser.

*Delta-Stak*®   Inclined-plate clarifier by Eimco Process Equipment Co.

**delta T**   Differential temperature.

*DeltΔ*™   Traveling water screen chain by Envirex, Inc.

*Delumper*®   Solids disintegrator and crusher products by Franklin Miller.

**demineralizing**   The process of removing minerals from water, most commonly through an ion exchange process.

*Demister*®   Mist eliminator by Otto H. York Co., Inc.

*Denite*®   Denitrification process using a granular media fixed-film reactor bed by TETRA Technologies, Inc.

**denitrification**   Biological process in which nitrate is converted to nitrogen and other gaseous endproducts.

*Denitri-Filt*™   Biological denitrification filter by Davis Water & Waste Industries, Inc.

*Densadeg*®   Thickener-clarifier unit with lamella zone and sludge recirculation by Infilco Degremont, Inc.

*Densator*®   High-density solids contact clarifier with primary and secondary mixing zones by Infilco Degremont, Inc.

**dense nonaqueous-phase liquids (DNAPL)**   Liquids that are immiscible in and denser than water.

**density**   The weight of a substance per unit of its volume.

**density current**   A flow of water through a larger body of water that retains its unmixed identity due to a difference in density.

*Densludge*   Digestion system with primary thickening unit by Dorr-Oliver, Inc.

*Dentrol*   Sludge density controller by Walker Process Equipment Co.

*Deox/2000*™   Dechlorination analyzer by Wallace & Tiernan, Inc.

**deoxyribonucleic acid (DNA)**  The macromolecule that contains the hereditary material vital to reproduction.

**Department of Energy (DOE)**  U.S. Federal agency responsible for research and development of energy technology.

**Department of Transportation (DOT)**  U.S. Federal agency responsible for regulating transport of hazardous and nonhazardous materials.

*Deplution*  Wastewater treatment products by Ralph B. Carter Co.

*Depolox®*  Chlorine, pH, and fluoride analyzer by Wallace & Tiernan, Inc.

**depth filtration**  Filtration classification for filters where solids are removed within the granular media.

*Depurator*  Induced air flotation unit by Envirotech Company.

*DESAL™*  Desalination system product line by Desalination Systems, Inc.

**desalination**  The process of removing dissolved salts from water.

**desalting**  See "desalination."

*DeSander®*  Hydrocyclone separator by Krebs Engineers.

**desert**  A region characterized by a climatic pattern where evaporation exceeds precipitation.

**desertification**  The process where the biological productivity of land is reduced, resulting in desert-like conditions.

**desiccant**  A substance capable of absorbing moisture, used as a drying agent.

**desorption**  The release of an adsorbed solute from an adsorbent.

**destruction and removal efficiency (DRE)**  An expression of hazardous waste incinerator efficiency stated as the percentage of incoming principal organic hazardous components destroyed during incineration.

*Destrux®* Heavy-duty rotary knife cutter to granulate scrap material by Franklin Miller.

**detention time** The period of time that a volume of liquid remains in a tank.

**detoxification** Treatment to remove a toxic material.

*Detritor* Grit removal unit with reciprocating raking mechanism by Dorr-Oliver, Inc.

**detritus** (1) Decaying organic matter such as root hairs, stems, and leaves usually found on the bottom of a water body. (2) Grit or fragments of rock or minerals.

**detritus tank** Square tank grit chamber incorporating a revolving rake to scrape settled grit to a sump for removal.

*Developure* Depth filter by Osmonics, Inc.

**dew** Water droplets that form on cool surfaces following condensation of atmospheric water vapor.

**dewatering lagoon** A lagoon constructed with a sand and underdrain bottom.

**dewatering table** See "belt thickener."

**dew point** The temperature to which air with a given concentration of water vapor must be cooled to result in condensation of the vapor.

*DFR* Dynamic fixed-film reactor by Schreiber Corp.

**DFT** See "dry film thickness."

**DI** See "deionization."

**dialysis** The separation of substances from solution on the basis of molecular size by diffusion through a semipermeable membrane.

*Diamite Series*™ Membrane cleaning liquids for removal of organic and inorganic foulants by King Lee Technologies.

*Diamond Gate*    Screenings press by Andritz-Ruthner, Inc. (Western Hemisphere) and Contra-Shear Engineering, Ltd.

*Diamond Seal*™    Metering gate by Tetra Technologies, Inc.

**diatom**    A unicellular algae with a yellowish brown color and siliceous shell.

**diatomaceous earth (DE)**    Skeletal deposits of diatoms used as filter aids and a filter medium.

**diatomaceous earth filter**    Water treatment filter that uses a layer of diatomaceous earth as the filter medium.

**diatomite**    See "diatomaceous earth."

**diazinon**    A common organophosphate insecticide.

*Diffusadome*®    Coarse-bubble air diffuser by Amwell, Inc.

*Diffusair*    Carbon dioxide diffuser system by Walker Process Equipment Co.

**diffused-air aeration**    The introduction of compressed air into water by means of submerged diffusers or nozzles.

**diffuser**    A porous plate or tube through which air, or another gas, is forced and divided into bubbles for diffusion in liquids.

*Diffuserator*    Carbon dioxide diffusion system by Walker Process Equipment Co.

*Digesdahl*    Digestion apparatus for lab sample preparation for crude protein and mineral analysis by Hach Company.

**digester**    A tank or vessel used for sludge digestion.

**digestion**    The biological oxidation of organic matter in sludge resulting in stabilization.

*Digichem*    Programmable titration analyzer by Ionics, Inc.

*Dijbo*    Hydraulically operated trash rake by Landustrie Sneek BV.

**dike**   An embankment or ridge of materials used to prevent the movement of liquid or sludges.

**dilatant**   Property of a liquid whose viscosity increases as agitation is increased.

**dilution**   (1) Lowering the concentration of a solution by adding more solvent. (2) A method of liquid disposal where wastewater or treated effluent is discharged into a stream or body of water.

**dilution factor**   The volumetric ratio of solvent to solute.

*Dimminutor®*   Open-channel comminutor by Franklin Miller.

**DIN**   Deutsche Industrie Normal.

**dioxin**   An aromatic halogenated hydrocarbon that is one of the most toxic compounds known.

*Dioxytrol*   Centrifugal oxygenation system by Hazleton Environmental Products, Inc.

*Dipair™*   Static tube aerator by Infilco Degremont, Inc.

*Diphonix™*   Ion exchange resins by Eichrom Industries, Inc.

*Di-Prime™*   Automatic trash pump by Goodwin Pumps.

**direct filtration**   Filtration process that does not include flocculation or sedimentation pretreatment.

*Director™*   Flow diversion baffle by Environetics, Inc.

*DirecTube*   Eductor tube anaerobic gas mixing unit by Walker Process Equipment Co.

*Discfuser®*   Coarse-bubble air diffuser by FMC Corp., MHS Division.

**discharge**   The release of any pollutant, by any means, to the environment.

*Discor*   Disc dryer by Andritz-Ruthner, Inc.

*Discostrainer*   Fine screening device by Hycor Corp.

*Discotherm*   Thermal sludge processor by LIST, Inc.

*Disc-Pak*   Cartridge filter by Alsop Engineering Co.

*Discreen®*   Rotating disc screening device by Ingersoll-Dresser Pump (U.S.) and H2O Waste-Tec (U.K.).

**discrete particle settling**   Phenomenon referring to sedimentation of particles in a suspension of low solids concentration.

**disc screen**   A screening device consisting of a circular disc fitted with wire mesh that rotates on a horizontal axis.

*Disc-Tube™*   Reverse osmosis system by Rochem Separation Systems.

*Di-sep*   Membrane filtration systems by Smith & Loveless, Inc.

**disinfection**   The selective destruction of disease-causing microbes through the application of chemicals or energy.

**disinfection byproducts (DBP)**   Byproducts that occur, or are anticipated to occur, from the addition of commonly used water treatment disinfectants, including chlorine, chloramine, chlorine dioxide, and ozone.

**dispersant**   An additive that prevents agglomeration of particulates, or is used to break up concentrations of organic matter, such as oil spills.

**dispersion rate**   A diffusion parameter of gas plumes or stack effluents.

*Disposable Waste Systems*   Former name of JWC Environmental.

**disposal**   The discharge, deposit, injection, dumping, spilling, leaking, or placing of any solid waste on land or water so that it may enter the environment or be emitted into the air.

*Disposorbs*   Self-contained disposable adsorption systems by Calgon Carbon Corp.

**dissolved air flotation (DAF)**   The clarification of flocculated material by contact with minute bubbles causing the air/floc mass to be buoyed to the surface, leaving behind a clarified water.

**dissolved nitrogen flotation (DNF)**  A variation of the dissolved air flotation process where nitrogen, rather than air, is used to assist in the removal of suspended solids.

**dissolved organic carbon (DOC)**  The fraction of total organic carbon that is dissolved in a water sample.

**dissolved oxygen (DO)**  The oxygen dissolved in a liquid.

**dissolved solids**  Solids in solution that cannot be removed by filtration. See "total dissolved solids."

**distill**  See "distillation."

**distillate**  A liquid product condensed from vapor during distillation.

**distillation**  The process of boiling a liquid solution, followed by condensation of the vapor, for the purpose of separating the solute from the solution.

**distributed control system (DCS)**  A collection of modules, each having a specific function, interconnected to carry out an integrated data acquisition and control operation.

**distributor**  See "rotary distributor."

**diurnal**  Occurring during a 24-hr period.

**diversion chamber**  A chamber used to divert all or part of a flow to various outlets.

**diversity index**  A mathematical expression that depicts the diversity of a species in quantitative terms.

*DLO*™  Dynamic light obscuration technique used in a particle monitor by Chemtrac Systems, Inc.

**DNA**  See "deoxyribonucleic acid."

**DNAPL**  See "dense nonaqueous-phase liquids."

**DNF**  See "dissolved nitrogen flotation."

**DO**  See "dissolved oxygen."

**DOAS**   Differential Optical Absorption Spectrometry.

**DOC**   See "dissolved organic carbon."

**doctor blade**   A scraping device used to remove or regulate the amount of material on a belt, roller, or other moving or rotating surface.

**DOE**   See "Department of Energy."

*Dokwed*   Back-cleaned bar screen by Hubert Stavoren BV.

**dolomite**   A natural mineral consisting of calcium carbonate and magnesium carbonate. Chemical formula is $CaMg(CO_3)_2$.

**dolomitic lime**   Lime containing 35–40% magnesium oxide.

**domestic wastewater**   Wastewater originating from sanitary conveniences in residential dwellings, office buildings, and institutions. Also called "sanitary wastewater."

*Don-Press*   Screw-type solids/screenings press by Dontech, Inc.

*DorrClone*   Grit classifier unit by Dorr-Oliver, Inc.

*Dorrco*   Brand name of Dorr-Oliver, Inc., products.

**dosage**   A specific quantity of a substance applied to a unit quantity of liquid to obtain a desired effect.

**dose-response**   The relationship between the dose of a pollutant and its effect on a biological system.

*Dosfolat®*   A stable, aqueous folic acid solution marketed by Bioprime.

**dosimeter**   An instrument used for measuring radiation dosage.

**dosing siphon**   A siphon that automatically discharges liquid onto a trickling filter bed or other wastewater treatment device.

**dosing tank**   A tank into which raw or partly treated wastewater is accumulated and held for subsequent discharge and treatment at a constant rate.

**DOT**   See "Department of Transportation."

*Double Dish*   Pressure filter underdrain by U.S. Filter Corp.

*Double Ditch*   Oxidation ditch wastewater treatment process by Krüger, Inc.

*DoubleGuard™*   Bar screen bar rack with two sets of offset bars by Envirex, Inc.

**double-suction pump**   A centrifugal pump with suction pipes connected to the casing from both sides.

*Dowex®*   Ion exchange resin by Dow Chemical Co.

**downcomer**   A pipe directed downward.

**DOX**   Dissolved organic halogen.

**DP**   Differential pressure.

*d-part®*   Waste conditioner by Medina Products.

**draft tube**   A centrally located vertical tube used to promote mixing in a sludge digester or aeration basin.

*Draft Tube Channel*   Oxidation ditch process formerly offered by Lightin.

*Drag-Star™*   Rectangular clarifier sludge removal device utilizing a cable drive and scraper carriage by Smith & Loveless, Inc.

**drag tank**   A rectangular sedimentation basin that uses a chain and flight collector mechanism to remove dense solids.

*Draimad*   Sludge dewatering system for small producers by Resi-Tech Division of Aero-Mod, Inc.

**drainage water**   Ground-, surface-, or stormwater collected by a drainage system and discharged into a natural waterway.

*Drain-Dri®*   Wet pit pump by Yeomans Chicago Corp.

**drain tile**   Short lengths of pipes laid in underground trenches to collect and carry away excess groundwater, or to discharge wastewater into the ground.

**DRE**   See "destruction and removal efficiency."

**dredge** To remove sediment or sludge from rivers or estuaries to maintain navigation channels.

*Dresser/Jeffrey* Screening equipment product line acquired by Jones and Attwood, Inc.

*Drewgard* Corrosion inhibitor by Ashland Chemical, Drew Division.

**drift** (1) Water lost from a cooling tower as mist or droplets entrained in the circulating air. (2) Pollutants entrained in a plant's stack discharge.

**drift barrier** An artificial barrier designed to catch driftwood or other floating material.

*Driftor*™ Drift eliminator by Kimre, Inc.

**drift test** A part of the emissions certification process in which the continuous emissions monitoring system must operate unattended for some period of time without the analyzers drifting out of calibration.

**drilling mud** A fluid, often containing bentonite, used to cool and lubricate a drilling bit and to remove cuttings from the bit and carry them to the well's surface.

**drinking water** Water safe for human consumption, or for use in the preparation of food or beverages, or for cleaning articles used in the preparation of food or beverages.

**drinking water equivalent level (DWEL)** The lifetime exposure level at which adverse health effects are not anticipated to occur, assuming 100% exposure from drinking water.

**Drinking Water Priority List (DWPL)** 1988 list of drinking water contaminants that may pose a health risk and warrant regulation.

**drip irrigation** A microirrigation water management technique used primarily for landscaping in which drips of water are emitted near the base of the plant.

**drip proof**   Designation for a motor enclosure with ventilating openings constructed so that drops of liquids or solids falling on the motor will not enter the unit directly or by running along an inwardly inclined surface.

**drought**   An extended period of dry weather, which, as a minimum, can result in a partial crop failure or an inability to meet normal water demands.

*Drumm-Scale™*   Drum weighing scale by Force Flow Equipment.

**drum pulverizer**   A rotating cylinder used to shred solid waste by the intermingling action of internal baffles acting on the wetted solid waste.

**drum screen**   A cylindrical screening device used to remove floatable and suspended solids from water or wastewater.

*Drumshear*   Rotating fine screen by Aer-O-Flo Environmental, Inc.

*Drumstik*   Feed tube for drum-mounted chemical feed system by Stranco, Inc.

*Dry-All*   Vacuum belt press by Baler Equipment Co.

**dry bulb temperature**   The air temperature measured by a conventional thermometer.

**drycleaning wastes**   Wastewater from laundry cleaning operations that use nonaqueous chemical solvents to clean fabrics.

**dry film thickness (DFT)**   Thickness of a dried paint or coating, usually expressed in mils.

**dry ice**   Solidified carbon dioxide, frequently used as a refrigerant, that vaporizes without passing through a liquid state.

**dry weather flow (DWF)**   The flow of wastewater in a sanitary sewer during dry weather. The sum of wastewater and dry weather infiltration.

**dry well** (1) A dry compartment in a pumping station where pumps are located. (2) A well that produces no water.

*DST*™ Defined Substrate Technology. Reagent system designed to promote growth of target microbes by Environetics, Inc.

*D Tech*™ Environmental field test kits by EM Industries, Inc.

*Dualator*® Gravity sand filter by Tonka Equipment Co.

**dual flow screen** A traveling water screen arranged in a channel so that water enters through both the ascending and descending wire mesh panels and exits through the center of the screen.

**dual media filter** Granular media filter utilizing two types of filter media, usually silica sand and anthracite.

*Dubl-Safe* Measuring system for brine regeneration of softening unit by Culligan International Corp.

*Duckbill*® Wastewater sampler by Markland Specialty Engineering.

**duckweed** See "Lemnaceae."

**duct** A tube or channel for gas or fluid flow.

*Dunkers* Flow balancing system by Munters.

*Duo-Clarifier* Wastewater treatment clarifier formerly offered by Dorr-Oliver, Inc.

*Duo-Deck* Floating cover for anaerobic digester tank by Envirex, Inc.

*Duo-Filter* Two-stage trickling filter formerly offered by Dorr-Oliver, Inc.

*Duo-Flo* Traveling mesh belt screen by Dontech, Inc.

*Duo-Pilot*® Coagulant control system using pilot filters by Wheelabrator Engineered Systems, Inc., Microfloc Products.

*DuoReel* Power cable reel system for clarifiers by Walker Process Equipment Co.

*DuoSparj*   Coarse-bubble diffuser by Walker Process Equipment Co.

*Duo-Vac*   Inclined screen unit by the former Bowser-Briggs Filtration Co.

*DuoVAL*   Automatically backwashed gravity sand filter by General Filter Co.

**duplex pump**   A reciprocating pump having two side-by-side cylinders and connected to the same suction and discharge lines.

**duplex stainless steel**   A high-strength stainless steel containing two forms of iron, typically austenite and ferrite.

*Dupont®*   Brand name of E.I. DuPont de Nemours, Inc.

*Dura-Disc*   Fine-bubble membrane diffuser by Wilfley Weber, Inc.

*Dura-Fuser™*   Aeration piping system by Davis Water & Waste Industries, Inc.

*Dura-Mix*   Mixer for aeration applications by Wilfley Weber, Inc.

*Dura-Trac™*   Sensor for streaming current transmitter by Chemtrac Systems, Inc.

*Durco*   Filtration equipment by Duriron Co., Inc.

*Durex®*   Roller chain material by Envirex, Inc.

*DuroFlow®*   Air blower by Gardner-Denver, division of Cooper Industries.

*Duroy*   Nonlubricated, metallic traveling water screen chain by Envirex, Inc.

**dust**   Fine-grained particles light enough to be suspended in air.

*Dustube®*   Fabric filter by Wheelabrator Clean Water Systems, Inc.

**DWEL**   See "drinking water equivalent level."

**DWF**   See "dry weather flow."

**DWPL**   See "Drinking Water Priority List."

*DWS*   Manufacturer of sewage grinders for JWC Environmental.

*Dynablend*™   Polymer blending and feeding system by Fluid Dynamics, Inc.

*DynaClear*®   Packaged water treatment system incorporating contact filtration by Parkson Corp.

*DynaCycle*®   Regenerative oxidation system by Monsanto Enviro-Chem Systems, Inc.

*DynaFloc Feedwell*   Clarifier feedwell design to promote mixing of flocculants by Dorr-Oliver, Inc.

*Dyna-Grind*   Screenings and sewage grinder by FMC Corp., MHS Division.

**dynamic head**   See "total dynamic head."

*DynaSand*®   Continuously backwashed moving bed sand filter by Parkson Corp.

*Dynasieve*™   Self-cleaning rotary fine screen by Andritz-Ruthner, Inc.

*Dynatherm*   In-vessel composting system by Compost Systems Co.

*Dynatrol*®   Measurement level switches and liquid measurement devices by Automation Products, Inc.

*DynaWave*®   Reverse jet scrubbing system by Monsanto Enviro-Chem Systems, Inc.

**dyne**   A unit of measurement equal to the force that imparts an acceleration of 1 cm/s$^2$ to a 1-gram mass.

*Dyneco*   Former bar screen manufacturer acquired by Parkson Corp.

**dysentery**   A disease of the gastrointestinal tract usually resulting from poor sanitary conditions and transmitted by contaminated food or water.

*Dystor*   Gas holder system for anaerobic digesters by Envirex, Inc.

**dystrophy**   A disorder caused by defective nutrition or metabolism.

# E

*E.A. Aerotor*   Packaged wastewater treatment plant by Lakeside Equipment Corp.

*EAD*   Electro-acoustical dewatering press by Ashbrook Corp.

**EAFD**   See "electric arc furnace dust."

*Earthtec®*   Low-pH algicide/bactericide by Earth Science Laboratories, Inc.

**easement**   A legal right to the use of land owned by others.

*EaseOut*   Pivoting air header and drop pipe arrangement by Walker Process Equipment Co.

*Eastern Econoline*   Mixer product line by EMI, Inc.

*EasyLogger*   Stormwater data logger by Omnidata International, Inc.

**ebb**   (1) The flowing back of water brought in by the tide. (2) To recede from a flooded state.

**ebb tide**   Tide occurring at the ebb period of tidal flow.

**EBCT**   Empty bed contact time.

*Echo-Lock*   Automatic compensating device for flow/level meters by Marsh-McBirney, Inc.

*ECI*   Environmental Conditioners, Inc., a former manufacturer of packaged wastewater treatment plants.

**ECL**   Electrochemiluminescence. A technology used for analyzing molecular and genetic material.

*Eclipse®*   Dispersion polymers by Calgon Corp.

*EcoCare™*   Odor eliminator by Nature Plus, Inc.

*Ecochoice*   Catalytic oxidation system for dissolved organics by Eco Purification Systems USA, Inc.

*Ecodenit*   Biological nitrate removal process using ion exchange technology by Biwater-OTV Ltd.

*EcoDry*   Sludge drying system by Andritz-Ruthner, Inc.

*Ecodyne*   Former name of Transunion product group now known as Graver Co.

*E. coli*   See *"Escherichia coli."*

*Ecolo-Chief*   Packaged wastewater treatment plants by Chief Industries, Inc.

*EcoLogic*™   Combination aeration and ultraviolet/ozonation system by Atlantic Ultraviolet Corp.

**ecology**   The relationship of living things to one another and their environment.

*Ecomachine*   Belt filter press by WesTech Engineering, Inc.

*Econ-Abator*®   Catalytic oxidation system by Wheelabrator Clean Air Systems, Inc.

*Econex*   Counterflow regeneration ion exchange by Ionics, Inc.

*Econ-NOx*™   Selective catalytic reduction system by Wheelabrator Clean Air Systems, Inc.

*Economixer*   Solid bowl decanter centrifuge by Centrisys Corp.

**economizer**   A heat exchanger in a furnace stack that transfers heat from the stack gas to the boiler feedwater.

**economy**   In thermal desalination, the ratio of pounds of distillate produced per 1000 Btu of energy input, or kilograms of distillate produced per 2326 kJ of energy input.

**economy-of-scale**   The reduction of unit capital cost realized as the size of the unit increases.

*Econopure*   Reverse osmosis system by Osmonics, Inc.

*Ecopure*   Volatile organic compound and particulate emission control product line by Dürr Environmental Division.

**ecorock** A hard, dense rock produced from the ash of incinerated sludge and municipal solid waste suitable for use as a road aggregate.

*Ecosorb®* Odor control neutralizer by Odor Management, Inc.

**ecosystem** The total community of living organisms, together with their physical and chemical environment.

*EcoVap™* Volatile organic compound control system by Wheelabrator Clean Air Systems, Inc.

**ECRA** Environmental Cleanup Responsibilities Act.

**ED** See "electrodialysis."

**EDAT** Environmental data acquisition telemetry.

**eddy** A vortex-like motion of a fluid running contrary to a main current.

*Eddyflow* High-rate upflow clarifier by Gravity Flow Systems, Inc.

*Edge Track* Drum filter cloth alignment mechanism by Eimco Process Equipment Co.

*EDI* Environmental Dynamics, Inc.

**EDR** See "electrodialysis reversal."

**EDS** European Desalination Society.

**EDTA** Ethylenediaminetetraacetic acid. A chelating agent.

*Eductogrit* Aerated grit chamber by Aerators, Inc.

*Edward & Jones* Filter press products marketed by Asdor.

**effect** One of several units of an evaporator, each of which operates at successively lower pressures.

**effective size** Method of characterizing filter sand where the effective size is equal to the sieve size, in millimeters, that will pass 10%, by weight, of the sand.

**effective stack height** The sum of the plume rise and physical stack height at which particulates in a stack emission begin to settle to the ground after discharge.

**effluent**   Partially or completely treated water or wastewater flowing out of a basin or treatment plant.

**EIA**   Environmental impact assessment.

*EimcoBelt®*   Continuous-belt vacuum filter by Eimco Process Equipment Co.

*EimcoMet*   Molded polypropylene components by Eimco Process Equipment Co.

**ejector**   A device that uses steam, air, or water under pressure to move another fluid by developing suction through the use of a venturi.

*Elasti-Liner®*   Containment liner material by KCC Corrosion Control Co.

*Elastol®*   Additive used to change properties of oil to enhance oil/water separation by Lemacon Techniek B.V.

**elastomer**   Synthetic material that is elastic or resilient and similar in structure, texture, and appearance to natural rubber.

*Elastox®*   Membrane air diffuser by Eimco Process Equipment Co.

*Elbac*   Wastewater bioaugmentation product by Eltech International Corp.

*Elbow Rake*   Hydraulically operated trash rake by the former Acme Engineering Co., Inc.

*Electraflote*   Sludge thickener using electrolysis-generated bubbles formerly offered by Ashbrook Corp.

**electric arc furnace dust (EAFD)**   A byproduct of the production of steel using electric arc furnaces, usually containing heavy metals.

**electrochemical corrosion**   Corrosion brought about through electrode reactions.

**electrodialysis (ED)**   The separation of a solution's ionic components through the use of semipermeable, ion-selective membranes operating in a direct current electric field.

**electrodialysis reversal (EDR)**   A variation of the electrodialysis process using electrode polarity reversal to automatically clean membrane surfaces.

**electrolysis**   The passage of electric current through an electrolyte, resulting in chemical changes caused by migration of positive ions toward the cathode and negative ions to the anode.

**electrolyte**   A substance that dissociates into two or more ions when it dissolves in water.

*Electromat*   Electrodialysis equipment by Ionics, Inc.

*Electromedia*®   Processed mineral filter media by Filtronics, Inc.

**electrometric titration**   An acid or base titration where a pH meter is used for measuring endpoints.

**electronic-grade water**   Water used in the production of microelectronic devices that meets D-19 standards of the American Society for Testing and Materials for resistivity, silica concentration, particle count, and other criteria.

**electron microscope**   A microscope that utilizes electromagnets as lenses and electrons instead of light rays to achieve a very high magnification.

**electrostatic precipitator (ESP)**   Air cleaning system that imparts an electrical charge to airborne particles so they can be removed by attraction to elements of opposite polarity.

**elevated tank**   A water storage reservoir supported by a column or tower.

*Elf/Anvar*   Oil/condensate coalescer-type oil/water separator by Graver Co.

**Elf Atochem** Former name of Atochem North America, Inc.

*El Niño* A climatic cycle resulting in warm, stormy weather in the Pacific caused by the warming of surface waters in the eastern Pacific Ocean.

*EloxMonitor*™ On-line COD monitor by Anatel Corp.

**elutriation** The process of washing sludge with water to remove organic and inorganic components to reduce chemical dosages required for additional treatment.

**elutriator** An extension in an evaporator vapor body to thicken the solids slurry to minimize the loss of liquor.

**EM** See "Enhanced Monitoring."

**EMC** Emission reduction credits.

*Emerzone*® Ozone generating systems by Emery-Trailigaz Ozone Co.

**EMI** Electromagnetic interference.

**emission** Gas-borne particles or pollutants released into the atmosphere.

**EMP** Electromagnetic pulse.

*EMR*™ Metal recovery system by Kinetico Engineered Systems, Inc.

**emulsifying agent** An agent that aids in creating and maintaining an emulsion by altering the surface charge of droplets to prevent their coalescence.

**emulsion** A heterogeneous mixture of two or more mutually insoluble liquids that would normally stratify according to their specific gravities.

**emulsion breaker** A demulsifying agent that breaks an emulsion by neutralizing the surface charge of the emulsified droplets to allow their coalescence.

**encapsulation** The complete enclosure of a waste in another material to isolate it from the external effects of air and water.

**encrustation**   A covering or crust or crust-like material on the surface of an object.

**endemic**   Restricted to a particular area or locality.

**endogenous respiration**   Bacterial growth phase in which microbes metabolize their own protoplasm without replacement due to low concentration of available food.

**endospore**   A bacterial spore formed within a cell and extremely resistant to heat and other harmful agents.

**endothermic**   A process or reaction that takes place with absorption of heat.

**endotoxin**   A toxin, or poisonous substance, present in bacteria that is released during cell lysis.

*Endurex*   Coarse-bubble diffuser by Parkson Corp.

*Enelco*   Former Environmental Elements Company water treatment product line acquired by Infilco Degremont, Inc.

*Energy Mix*   Rapid mix unit by Walker Process Equipment Co.

**energy recovery**   The retrieval of waste energy for some beneficial use.

**Enhanced Monitoring (EM)**   Clean Air Act Amendment requirement for facilities to monitor emissions to certify compliance with permitted levels.

**Enning ESD**   Enhanced sludge digestion egg-shaped anaerobic digesters by CBI-Walker, Inc. (U.S. licensee) and Enning (German licensor).

**enteric bacteria**   Bacteria that inhabit the gastrointestinal tract of warm-blooded animals.

*Enterolert*™   Reagent for enterococci detection by IDEXX Laboratories, Inc.

**enterotoxin**   A toxin or microbe that causes dysfunction in the human gastrointestinal tract.

*Enterprise*   Floating aspirating aerator by Air-O-Lator Corp.

*Enterprise™*   Regenerative thermal oxidizer by Grace TEC Systems.

**enthalpy**   The total heat content of a liquid, vapor, or body.

**entrainment**   (1) The incorporation of small organisms, including the eggs and larvae of fish and shellfish, into an intake system. (2) The carryover of droplets of water with vapor produced during evaporation.

**entrainment separator**   See "mist eliminator."

**entropy**   A measure of unavailable energy of an isolated thermodynamic system.

*Envessel Pasteurization™*   Lime stabilization and pasteurization sludge treatment process to further reduce pathogens by RDP Company.

*Enviro-Blend™*   Specialty chemicals for heavy metal waste treatment by American Minerals, Inc.

*Envirodisc*   Rotating biological contactor by Walker Process Equipment Co.

*Envirofab*   Former manufacturer of wastewater treatment equipment.

*ENVIROFirst™*   Sodium carbonate peroxyhydrate granules by Solvay Interox.

*EnviroGard™*   Chemical screening test kits from Millipore.

*Enviromat*   Wastewater treatment systems by Ionics, Inc.

**environment**   Water, air, and land, and the interrelationship that exists among and between water, air, and land and all living things.

*Environmental Elements*   Equipment manufacturer whose Water Treatment Division product line was acquired by Infilco Degremont, Inc.

**environmental impact assessment (EIA)**   A method of analysis that attempts to predict probable repercussions of a proposed development on the social and physical environment of the surrounding area.

**environmental impact statement**   A detailed written report that identifies and analyzes the environmental impact of a proposed action.

**Environmental Protection Agency (EPA)**   U.S. agency with primary responsibility for enforcing federal environmental laws.

*Enviropac*   Rotating biological contactor by Walker Process Equipment Co.

*Enviropax*   Tube settlers by Enviropax, Inc.

*Enviropress*   Piston-type screenings press by Environmental Engineering Ltd.

*Enviro-Seal*™   Valve packing system to prevent fugitive air emissions by Fisher Controls International, Inc.

*Envirosorb*™   Oil spill absorbent by Geosource, Ltd.

*Envirovalve*   Telescopic valve by EnviroQuip Corp.

**enzyme**   Organic catalyst that converts a substrate or nutrient to a form that can be transported into a cell.

**EPA**   See "Environmental Protection Agency."

**EPC**   Engineer, procure, and construct.

*EPCO*™   Rotating biological contactors by U.S. Filter Corp.

*EPIC*™   Enhanced polymer control system by Norchem Industries.

**epidemic**   An outbreak of disease affecting many people at one time.

**epidemiology**   The study of the incidence, distribution, and control of disease in a population.

**epilimnion**   The upper layer in a stratified lake that results from varying water densities.

**epm** See "equivalents per million."

**EPRI** Electric Power Research Institute.

**Epsom salt** Hydrated magnesium sulfate, having cathartic properties; also used in leather tanning and textile dyeing. Chemical formula is $MgSO_4 \cdot 7H_2O$.

**equalization** The process of dampening hydraulic or organic variations in a flow so that nearly constant conditions can be achieved.

**equalization basin** A basin or tank used for flow equalization.

**equivalents per million (epm)** Ionic concentration determined by dividing an ion's concentration in parts per million by its equivalent weight.

**equivalent weight** The weight of a compound that contains 1 gram atom of available hydrogen or its chemical equivalent, determined by dividing the molecular weight of a solute by the number of hydrogen or hydroxyl ions in the undissolved compound.

**ERDA** Energy Research and Development Administration.

**erosion** Wearing away of land by running water, waves, wind, or glacial activity.

**erosion corrosion** An attack on a material consisting of simultaneous erosion and corrosion through the effect of a rapidly flowing fluid.

**ESA** Endangered Species Act.

**escarpment** A line of steep slopes or cliffs caused by erosion or faulting.

*Escherichia coli* **(E. coli)** Coliform bacteria of fecal origin used as an indicator organism in the determination of wastewater pollution.

*ESD*™ Enhanced sludge digestion egg-shaped anaerobic digesters by CBI-Walker, Inc. (U.S. licensee) and Enning (German licensor).

*ESD-Hormel*   Former manufacturer of products offered by U.S. Filter Corp.

**ESP**   See "electrostatic precipitator."

*ESP®*   Sludge drying and pelletization system by Wheelabrator Clean Water Systems, Inc., Bio Gro Division.

*ESPA™*   Low-pressure polyamide membrane products by Hydranautics.

*ESSI*   EnviroSystems Supply, Inc.

**estuary**   A semi-enclosed coastal water body at the mouth of a river in which the river's current meets the sea's tide.

**ESWTR**   Enhanced Surface Water Treatment Rule.

**ethanol**   An inflammable organic compound formed during the fermentation of sugars. Chemical formula is $C_2H_5OH$.

**EU**   Endotoxin units.

**euphotic zone**   The upper layer of water in a natural water body through which sunlight can penetrate.

*Euroform*   Mist eliminators by Munters.

**eutectic**   Easily melted.

**eutrophication**   Enrichment of water, causing excessive growth of aquatic plants and eventual deoxygenation of the water body.

**eutrophic lake**   A lake with an abundant supply of nutrients, excessive growth of floating algae, and an anaerobic hypolimnion.

*Eva*   Back-raked bar screen by Brackett Green, Ltd.

**evaporation**   The process by which water is converted to a vapor that can be condensed.

**evaporation pond**   A natural or artificial pond used to convert solar energy to heat to accomplish evaporation.

**evaporation rate**   The mass quantity of water evaporated from a specified water surface per unit of time.

**evaporator** A device used to heat water to create a phase change from the liquid to the vapor phase.

**evaporite** A mineral produced as a result of evaporation.

**evapotranspiration** Water withdrawn from the soil by evaporation and plant transpiration.

**evapotranspiration treatment system** A wastewater treatment system utilizing surface evaporation and plant transpiration.

*EVT* Belt filter press by Eimco Process Equipment Co.

*Eweson®* Compartmentalized rotary digester by Bedminster Bioconversion Corp.

**EWPCA** European Water Pollution Control Association.

*Excel®* High-charge cationic flocculant by Cytec Industries, Inc.

**excess lime-soda softening** The process of feeding excess lime and soda ash in addition to that required for lime-soda softening to further reduce hardness. Also called "railway softening."

**exchange capacity** An ion exchanger unit's limited capacity for storage of ions.

**excyst** To emerge from a cyst.

**exhaustion** That condition that results when activated carbon, ion exchange resin, or other absorbents have depleted their capacity by using all available sites.

**exothermic** A process or reaction that is accompanied by the evolution of heat.

*Exotox®* Multigas detector by Neotronics.

**expanded metal** An open metal network produced by stamping or perforating sheet metal.

**expansion joint** A joint installed in a structure to allow for thermal expansion or contraction.

**explosion proof (XP)** Designation for a motor or electrical enclosure designed to withstand a gas or vapor explosion within the unit, and to prevent ignition of gas or vapor surrounding the unit by sparks, flashes, or explosions within the unit.

*ExpressClean*™   Reverse osmosis cleaning service by Coster Engineering.

*Expressor*   Belt filter press by Eimco Process Equipment Co.

**extended aeration process**   A variation of the activated sludge process with an increased detention time to allow endogenous respiration to occur.

*Extendor*   Detention tank for polymer mixing systems by Semblex, Inc.

**extraction steam**   Steam removed from a turbine at a pressure higher than the lowest pressure achieved in the turbine.

*Extractor*   Horizontal belt press by Eimco Process Equipment Co.

*Extractoveyor*   Composted sludge conveyor system by Compost Systems Co.

*Extreme Duty*™   Sludge mixer by WesTech.

*E-Z*™   Batch-type centrifuge by Western States Machine Company.

*E-Z Tray*®   Air stripper by QED Environmental Systems.

# F

**facultative bacteria**    Microbes with the ability to survive with or without the presence of dissolved oxygen.

**facultative lagoon**    A lagoon or pond in which stabilization of wastewater occurs as a result of aerobic, anaerobic, and facultative bacteria.

**facultative ponds**    See "facultative lagoons."

*Fairfield*    In-vessel composting system by Compost Systems Co.

**falaj**    System of surface or subsurface channeling of water fed by wells or springs to provide a community water supply.

**fall**    A sudden change in the water surface elevation.

**falling film evaporator**    An evaporator with vertical heat transfer surfaces where liquor falling down the surfaces is heated by steam condensing on the other side of the surface.

**fallout**    Radioactive debris that settles to earth after a nuclear explosion.

*Fallova*™    Sewage shredder by International Shredder, Inc.

**famine**    An acute food shortage in an area that leads to malnutrition and starvation.

**FAO**    Food and Agriculture Organization.

**Faraday's law**    The amount of chemical change in an electrolysis process is proportional to the electrical charge passed.

*Farm Gas*    Former equipment manufacturer now part of Rosewater Engineering Ltd.

*Fasflo*   Floating oil skimmer by Vikoma International.

*FAST®*   Fixed activated sludge wastewater treatment plant by Scienco/Fast Systems.

*Fastek™*   Thin-layer composite reverse osmosis membrane by Osmonics.

**fauna**   The animal life of an area or region.

*Favair®*   Dissolved-air flotation product by U.S. Filter Corp.

**FBC**   Fluidized bed combustion.

**FBS**   Fine-bar screen by Jones and Attwood, Inc.

**FC/FS**   Ratio of fecal coliform to fecal streptococci; it indicates if wastewater contamination results from animal or human wastes.

**FDA**   Food and Drug Administration.

*$Fe^3$®*   Liquid ferric sulfate by FE3, Inc.

**fecal coliform**   Coliforms present in the feces of warm-blooded animals.

*Fecascrew*   Screenings screw press by Hydropress Wallender & Co.

*Fecawash*   Screenings washing and conveying unit by Hydropress Wallender & Co.

**feces**   Excrement of humans and animals.

**Federal Insecticide, Fungicide and Rodenticide Act (FIFRA)**   1972 federal law requiring toxicity testing and registration of pesticides.

*Federal Register*   U.S. government daily publication detailing federal business including proposed and final rules and laws.

**Federal Water Pollution Control Act (FWPCA)**   U.S. water control legislation of 1972 amended by the Clean Water Act of 1977.

**feedlot wastes**   Solid and liquid wastes from a facility where cattle or other animals are raised for market.

*Feedpac*  Coagulant feed system by Nalco Chemical Co.

**feedwater heater**  Heat exchangers in which boiler feedwater is preheated by steam extracted from a turbine.

**FEMA**  Federal Emergency Management Agency.

**F.E.M.S.®**  Relational database fugitive emissions management system by EnviroMetrics.

**Fenton's reagent**  A mixture of iron and hydrogen peroxide used to chemically oxidize or degrade toxic organic chemicals in soil.

*Feripac*  Water treatment plant for iron and manganese removal by Vulcan Industries, Inc.

**fermentation**  The anaerobic conversion of organic matter to carbon dioxide, methane, and other low-molecular-weight compounds.

**ferric**  Relating to or containing iron that is trivalent or in a higher state of oxidation.

**ferric chloride**  A commonly used coagulant. Chemical formula is $FeCl_3$.

*FerriClear®*  Ferric sulfate product by Eaglebrook, Inc.

**ferric sulfate**  A commonly used coagulant. Chemical formula is $Fe_2SO_4$.

*Ferri-Floc*  Ferric sulfate by Boliden Intertrade, Inc.

*Ferrosand®*  Granular filter media used for removal of iron and manganese by Hungerford and Terry, Inc.

**ferrous**  Relating to or containing iron that is divalent or in a lower state of oxidation.

**ferrous sulfate**  A commonly used coagulant. Chemical formula is $FeSO_4$.

*Ferrover*  Reagent chemicals used primarily for iron analysis by Hach Company.

*Ferrozine*  Spectrophotometric reagents for iron and iron compounds by Hach Company.

*Ferr-X*  Iron removal process by Aquatrol Ferr-X Corp.

**fertilizer**   Materials, usually containing nitrogen and phosphorus, that are added to soil to provide essential nutrients for plant growth.

**FGD**   See "flue gas desulfurization."

**FGH**   See "flue gas humidification."

*Fibercone Press*   Cone-type dewatering press by Black Clawson, Shartles Division.

*FiberFlo™*   Cartridge filters by Fibercor.

**FIFRA**   See "Federal Insecticide, Fungicide and Rodenticide Act."

**filamentous growth**   Hair-like biological growth of some species of bacteria, algae, and fungi that results in poor sludge settling.

**filamentous sludge**   Sludge characterized by excessive growth of filamentous bacteria that results in poor sludge settling.

*Filawound®*   Pressure vessel for reverse osmosis membranes by Spaulding Composites.

**fill-and-draw**   Treatment process where a vessel is filled, the reaction occurs, and the contents are withdrawn; typical of "sequencing batch reactors."

*FilmShear*   Coarse- and fine-bubble diffusers by Aerators, Inc.

*Filmtec®*   Reverse osmosis membranes by Dow Chemical Co.

*Filox®*   Iron, hydrogen sulfide, and manganese removal media by Matt-Son, Inc.

*FiltaBand*   Continuous self-cleaning fine screen by Longwood Engineering Co., Ltd.

**filter**   A device utilizing a granular material, woven cloth, or other medium to remove suspended solids from water, wastewater, or air.

*Filter AG®*   Granular filter media used to adsorb tastes and odors by Clack Corp.

**filter aid**  A polymer or other material added to improve the effectiveness of the filtration process.

**filter bottom**  See "underdrain."

*Filter Cel®*  Diatomaceous earth filter media by Celite Corp.

**filter cloth**  Cloth used as the filter medium on a vacuum filter.

**filter cycle**  The filter operating time between backwashes. Also called "filter run."

**filter fly**  See "Psychoda flies."

**filter gallery**  A passageway to provide access for installation and maintenance of underground filter pipes and valves.

*Filterite®*  Cartridge filter by Memtec America Corp.

**filter loading, hydraulic**  The volume of liquid applied per unit area of the filter bed per day.

**filter loading, organic**  The pounds of biochemical oxygen demand applied per unit area of the filter bed per day.

*Filtermate*  Filtration coagulant by Argo Scientific.

*FilterNet™*  Net fabric composite by SLT North America, Inc.

*Filterpak*  Plastic biological filter media by Mass Transfer, Inc.

*Filter-Pak™*  Gravity sand filter by Graver Co.

**filter press**  A sludge dewatering device where water is forced from the sludge under high pressure.

**filter run**  See "filter cycle."

*FilterSil™*  Filtration sands by Unimin Corp.

**filter-to-waste**  Filter operational procedure in which the filtrate produced immediately after backwash is wasted.

*Filtomat®*  Self-cleaning cooling water filters by Orival, Inc.

*Filtra-Matic™*   Pressure leaf filter by U.S. Filter Corp.

*FiltraPak*   Packaged waste treatment and liquid-solids separation unit by Diagenex, Inc.

*Filtrasorb®*   Granular activated carbon adsorption manufactured by Calgon Carbon Corp.

**filtrate**   Liquid remaining after removal of solids through filtration.

**filtration rate**   A measurement of the volume of water applied to a filter per unit of surface area in a stated period of time.

*Filtroba®*   Helical pressure filter element by Ketema, Inc.

*Filtromatic*   Traveling bridge filter by Biwater Treatment Ltd.

*Filtros®*   Fine-bubble diffusers by Ferro Corp.

**final clarifier**   See "secondary clarifier."

**final closure**   The closure of all hazardous waste management units at a facility in accordance with all applicable requirements.

**final cover**   Cover material that is applied upon closure of a landfill and is permanently exposed to the surface.

**final effluent**   The effluent from the final unit treatment process at a wastewater treatment plant.

*FineAir*   Ceramic fine-bubble diffuser by Parkson Corp.

**fine-bubble aeration**   Diffused aeration by means of fine bubbles having high oxygen transfer rates.

**fines**   Particles at the lower end of a range of particle sizes.

**fine sand**   Sand particles with diameters that usually range from 0.3 to 0.6 mm.

**fine screen**   A screening device usually having openings smaller than 6 mm.

**firetube boiler**   Boiler in which the flame and hot gases are confined within tubes arranged in a bundle within a water drum.

**first-order reaction** A reaction in which the rate of change is directly proportional to the first power of the concentration of the reactant.

**first-stage BOD** See "carbonaceous biochemical oxygen demand."

**fisheyes** A condition resulting from the improper mixing of dry polymer and water which results in the formation of lumps of undispersed polymer that resemble fisheyes.

**fish ladder** A structure that permits fish to bypass a dam, using a series of baffled chambers installed at progressively lower elevations to provide a velocity against which fish can more easily swim.

**fish screen** (1) A screen at the head of an intake channel to prevent fish from entering. (2) A traveling water screen modified to remove impinged fish and return them to the water body.

**fission** (1) The splitting of the nucleus of an atom into two or more nuclei with a concurrent release of energy. (2) A form of asexual reproduction where the parent organism splits into two independent organisms.

*Fitch Feedwell* Clarifier feedwell with three horizontal chambers by Dorr-Oliver, Inc.

**fixation** The stabilization or solidification of a waste material by involving it in the formation of a stable solid derivative.

**fixed-bed porosity** The ratio of void volume to total bed volume of a granular media filter.

**fixed cover** Stationary anaerobic digester cover that allows a constant digester tank volume to be maintained.

**fixed-film process** Biological wastewater treatment process where the microbes responsible for conversion of the organic matter in wastewater are attached to an inert medium such as rock or plastic materials. Also called "attached growth process."

**fixed matter**   See "fixed suspended solids."

**fixed suspended solids**   Inorganic content of suspended solids in a water or wastewater sample, determined after heating the sample to 600°C.

**fjord**   A narrow inlet of the sea bordered by cliffs or highlands. Also "fiord."

**FKC America**   Former name of American Screw Press, Inc.

**flagellates**   Microorganisms that move by the whipping action of tail-like projections called a flagella.

**flammable**   Easily set on fire.

**flange**   A projecting rim or edge used for attachment with another object.

**flap valve**   A valve that is hinged on one edge so that it opens in the direction of normal flow and closes with flow reversal.

**flash**   The portion of a fluid converted to vapor when its pressure is reduced below the saturation pressure.

**flash dryer**   Sludge drying process that involves pulverizing sludge in a cage mill or by an atomized suspension technique in the presence of hot gases.

**flash evaporator**   A distillation device where saline water is vaporized in a vessel under vacuum through pressure reduction.

**flashing**   The process of vaporizing a fluid by pressure reduction rather than temperature elevation.

**flash mixing**   Motor-driven stirring devices designed to disperse coagulants or other chemicals instantly, prior to flocculation.

**flash point**   The temperature at which a substance ignites.

**Flashvap**   Flash evaporator by Licon, Inc.

**flavor profile analysis**   An analysis profiling the matrix of odors in a water sample.

**Fletcher**   Filtration division by Edwards & Jones, Inc.

*FlexAir*™   Fine-pore membrane diffuser by Environmental Dynamics, Inc.

*Flex-A-Tube*®   Fine-bubble aeration diffuser by Parkson Corp.

*FlexDisc*   Fine-bubble membrane diffuser by EnviroQuip International.

*FlexDome*   Fine-bubble membrane diffuser by EnviroQuip International Inc.

*Flexi-Fabric*   Filter press filter fabric by Eimco Process Equipment Co.

*Flexiflo*   High-speed surface aerator by Aerators, Inc.

*Flexi Jet*   Air-sparging power mixer by Aerators, Inc.

*Flexi-Jet*   Spray nozzles for sand filter rotary surface skimmer by F.B. Leopold Co.

*Flex-i-liner*®   Sealless self-priming rotary pumps by Vanton Pump & Equipment Corp.

*Fleximix*   High-speed surface aerator by Aerators, Inc.

*Flexishaft*™   Progressing cavity pump by MGI Pumps, Inc.

*FlexKlear*   Inclined-plate settler by Eimco Process Equipment Co.

*FlexKleen*™   Filter underdrain nozzles by Eimco Process Equipment Co.

*FlexLine*™   Nonbuoyant tubular diffuser by EnviroQuip International.

*Flexmate*®   Skid-mounted water purification vessel by U.S. Filter Corp.

*Flexofuser*®   Fine-bubble diffuser with tube body and fine sheath by FMC Corp., MHS Division.

*Flexoplate*™   Membrane-type medium-bubble diffuser by FMC Corp., MHS Division.

*FlexRO*™   Skid-mounted reverse osmosis system by U.S. Filter Corp.

*FlexRO Mobile®*  Mobile, skid-mounted reverse osmosis systems by U.S. Filter Corp.

*Flexscour*™  Filter underdrain for simultaneous air-water backwash by Eimco Process Equipment Co.

**flight**  (1) The horizontal scraper on a rectangular sludge collector. (2) The helical blade on a screw pump.

*Flight Guide*  Nonmetallic wear shoes for sludge collector flights by Trusty Cook, Inc.

*Flint Rim*  Cast sprocket with hardened, chill rim by FMC Corp., MHS Division.

*FloatAll*  Dissolved-air flotation system by Walker Process Equipment Co.

**floating cover**  Anaerobic sludge digester tank cover that is free to move up or down to change the total internal capacity of the tank in response to sludge additions or withdrawals.

**floating gas holder**  Anaerobic sludge digester tank cover that floats on a cushion of gas and moves up or down to change the total internal capacity of the tank in response to changes in gas volume.

**float switch**  An electrical or pneumatic switch operated by a float in response to changing liquid levels.

*Float-Treat®*  Dissolved-air flotation system by Envirex, Inc.

*Flo-Buster*™  Organic solids agitator/separator by Enviro-Care.

**floc**  Small, gelatinous masses formed in water by adding a coagulant, or in wastewater through biological activity.

*Floc Barrier®*  Inclined settling tubes for clarifiers by Graver Co.

**flocculants**  Organic polyelectrolytes, used alone or with metal salts that are used as coagulants.

**flocculant settling** Sedimentation of particles in a dilute suspension as they coalesce or flocculate.

**flocculation** Gentle stirring or agitation to accelerate the agglomeration of particles to enhance sedimentation or flotation.

**flocculator** A device used to enhance the formation of floc through gentle stirring or mixing.

*Flocide®* Sanitizer additive by FMC Corp., Process Additives Division.

*Floclean* Membrane cleaning product by FMC Corp., Process Additives Division.

*Floclear* Solids contact clarifier by Aerators, Inc.

*Flocon®* Reverse osmosis feedwater additive by FMC Corp., Process Additives Division.

*Flo-Conveyor™* Screenings conveyor by Enviro-Care Co.

*FloCor* Crossflow trickling filter media by the former Gray Engineering Co.

*Floc-Pac™* Compact flocculant packages by American Cyanamid Co.

*Flocpress* Belt filter press formerly offered by Infilco Degremont, Inc.

*Flocsettler* Combination flocculator/clarifier by Amwell, Inc.

*Flocsillator* Horizontal oscillating flocculator by Eimco Process Equipment Co.

*FlocTreator* Coagulation and flocculation unit by PWT Americas.

*Floctrol* Multiple-stage, horizontal shaft, paddle flocculators by Envirex, Inc.

*Flofilter™* Water treatment system combining flotation and filtration by Purac Engineering, Inc.

*Flo-Lift™* Vertical screenings lift device by Enviro-Care Co.

*Flo-Line*   Fine-mesh screening machine by Derrick Corp.

*FloMag*™   Magnesium-based granular, powder, and slurry water treatment products by Martin Marietta Specialties, Inc.

*FloMaker*™   Submersible mixer by ITT Flygt Corp.

*Flo-Mate*   Portable flow meter by Marsh-McBirney, Inc.

**flood plain**   The lowland area adjoining inland and coastal waters subject to a 1% or greater chance of flooding in a given year.

*Flo-Poke*   Portable flow rate instrument by Isco, Inc.

**flora**   Plants and plant life of a particular region or period.

*Flo-Screen*™   Reciprocating rake bar screen by Enviro-Care Co.

**flotation**   A treatment process in which gas bubbles are introduced to a water and attach to solid particles creating bubble-solid agglomerates that float to the surface, from which they are removed.

**flotation thickening**   Sludge thickening by means of dissolved-air flotation.

*Flo-Tote*   Computerized open-channel flow meter by Marsh-McBirney, Inc.

**flotsam**   Floating debris resulting from human activity.

*Flo-Ware*   Data logging flow meter software by Marsh-McBirney, Inc.

**flow control valve**   A device that controls the rate of fluid flow.

*Flow Logger*   Flow measuring device by Isco, Inc.

*Flowminutor*™   Horizontal shaft comminutor by Enviro-Care Co.

**flow rate**   The volume or mass of a gas, liquid, or solid material that passes some point in a stated period of time.

*FlowSorb*™   Granular activated carbon canisters by Calgon Carbon Corp.

**flow splitter**  A chamber that divides incoming flow into two or more streams.

*Flowtrex*  Pleated cartridge filter by Osmonics, Inc.

**flue**  Any passage designed to carry combustion gases and entrained particulates.

**flue gas**  The gases and smoke that are released from an incinerator's chimney.

**flue gas desulfurization (FGD)**  The process of removing sulfur dioxide from exhaust gas, usually by means of a wet scrubbing process.

**flue gas humidification (FGH)**  A process to control $SO_2$ emissions by spraying a water/air mixture into flue gas.

*FluePac™*  Powdered activated carbon by Calgon Carbon Corp.

**fluid**  Any material or substance that flows or moves whether in a semisolid, liquid, sludge, or gaseous form or state.

*Fluidactor*  Fluid bed sludge incinerator formerly offered by Walker Process Equipment Co.

*Fluid Dynamics®*  Engineered filter system by Memtec America Corp.

**fluidization**  The upward flow of a gas or fluid through a granular bed at sufficient velocity to suspend the grains.

**fluidized bed combustion**  A method of burning particulate fuel, such as powdered coal, where the fuel is burned after injection into a rapidly moving gas stream.

**fluidized bed furnace**  An incinerator or furnace used to incinerate sludge by passing heated air upward through a fluidized sand bed.

*Fluidizer-Minor*  Flash dryer by Centrico, Inc.

**flume**  A channel used to carry water.

*Fluorapid*  Process for increasing clarifier rise rates by Biwater-OTV Ltd.

*Fluoretrack*   Liquid water tracing dye by Formulabs, Inc.

**fluoridation**   The addition of fluoride to drinking water to aid in the prevention of tooth decay.

**fluorimeter**   An instrument used to measure the amount of fluorescent materials, dyes, or aromatic hydrocarbons in water.

**fluorspar**   Common source of commercially available fluoride compounds; it may be used as a direct source of fluoride for water fluoridation.

*Fluosolids*   Fluid bed reactor for sludge disposal by Dorr-Oliver, Inc.

*Flush-Kleen®*   Nonclogging sewage ejector pump by Yeomans Chicago Corp.

**flux**   (1) Flow rate per unit area. (2) Heat transfer rate per unit area.

**fly ash**   The noncombustible particles in flue gas.

*Flygt*   Pump and mixer products by ITT Flygt Corp.

**F/M**   See "food-to-microorganism ratio."

**FmHA**   Farmers Home Administration.

*Foam Ban*™   Foam control additive by Ultra Additives, Inc.

*Foamtrol*™   Foam control additive by Ultra Additives, Inc.

**fodder crop**   A crop grown principally for animal food.

**FOG**   Fats, oils, and grease.

**fomite**   Objects that have been contaminated by pathogens from a diseased person and may serve in their transmission.

*Font'n Aire®*   Floating decorative aerator/fountain by Air-O-Lator Corp.

**food chain**   A feeding hierarchy in which food energy is passed from primary producers (plants) to primary consumers (herbivores) to secondary consumers (carnivores).

**food chain crops** Crops grown for human consumption, and crops grown for feed for animals whose products are used for human consumption.

**food crop** Crops grown primarily for human consumption.

**food poisoning** A gastrointestinal disorder caused by bacteria or their toxic products, occurring after consumption of contaminated food.

**food-to-microorganism ratio (F/M)** The ratio of the influent biochemical oxygen demand to the volatile suspended solids concentration in a wastewater treatment aeration tank.

**food waste** Organic residues generated by the handling, storage, preparation, cooking, and serving of food.

**food web** The complex interconnected network of food chains in a community.

**foot-pound** Unit of measure of the work performed by a force of 1 lb acting through a distance of 1 ft.

**forage crop** Crops that can be used as feed by domestic animals, by being grazed or cut for hay.

*Forager*™ Sponge containing a bonded polymer, used to selectively remove heavy metals from water by Dynaphore, Inc.

**force main** The pressurized pipeline between the pumped discharge at a water or wastewater pumping station and a point of gravity flow.

**forced circulation evaporator** An evaporator in which circulation is maintained by pumping the liquid through the heating element with relatively low evaporation per pass.

**forced draft deaerator** Device to remove dissolved gases from solution by blowing an air stream through a packed column countercurrent to downflowing water.

**forebay** A reservoir at the end of a pipeline or channel.

**formazin turbidity unit (FTU)**   Unit of measure used in turbidity measurement based on a known chemical reaction that produces insoluble particulates of uniform size.

*Fossil Filter*™   Stormwater runoff filter by KriStar Enterprises.

**fossil fuel**   Natural gas, petroleum, coal, and any form of solid liquid or gaseous fuel derived from such materials for the purpose of creating useful heat.

**fossil water**   See "connate water."

*Fotovap*   Evaporator to treat photoprocessing lab wastewater by Licon, Inc.

**fouling**   Condition caused when bacterial growth, colloidal material, or scale forms a deposit on a filter, membrane, or heat transfer surface.

**fouling factor**   A design criteria used to allow for some variation of equipment performance due to fouling.

*4-Beam*™   Turbidity and suspended solids measuring instrument by BTG, Inc.

**four 9s DRE**   Common term for an incinerator "destruction and removal efficiency" of 99.99%.

**Fourier transform infrared spectrometry (FTIR)**   Continuous emissions monitoring system used to identify organic and inorganic compounds in liquids, solids, and gases.

*Fox-Pac*   Sanitation system for marine applications by Red Fox Environmental, Inc.

**FPA**   See "flavor profile analysis."

**fpm**   Feet per minute.

**fps**   Feet per second.

**fractionation**   A distillation method used to separate a mixture of several volatile components of different boiling points in successive stages, with each stage removing some proportion of one component.

**frazil ice**    Granular or spike-shaped ice crystals that form in supercooled water that is too turbulent to permit coagulation into sheet ice.

**freeboard**    The vertical distance between the normal maximum liquid level in a basin and the top of the basin that is provided so that waves and other liquid movements will not overflow the basin.

**free chlorine**    The amount of chlorine available as dissolved gas, hypochlorous acid, or hypochlorite ion.

**free chlorine residual**    Portion of the total residual chlorine remaining at the end of a specific contact time which will react as hypochlorous acid or hypochlorite ion.

*Free-Flow*™    Ceramic diffuser plates for fine-bubble aeration by Davis Water & Waste Industries, Inc.

**free liquids**    Liquids that readily separate from the solid portion of a waste under ambient temperature and conditions.

**free oil**    Nonemulsified oil that separates from water, usually in 5 min or less.

**free settling**    The settling of discrete, nonflocculant particles in a dilute suspension.

*Free-Slide*    Wire mesh basket configuration for traveling water screens by Envirex, Inc.

**free water knockout (FWKO)**    Gravity separation vessel used in an oil field to separate produced water from oil.

**freeze distillation**    Production of distillate by freezing a saline solution and washing salts from the pure water crystals prior to melting.

*Fre-Flo*    Extruded cement underdrain for a gravity filter by Infilco Degremont, Inc.

*Freon*®    Refrigerant compound by E. I. du Pont de Nemours.

**fresh water**   Water that usually contains less than 1000 mg/L of dissolved solids.

**Freundlich isotherm**   Graphical representation of data related to the removal of colloidal matter from water as a result of adsorption.

**FREX**   Freon extractable oil and grease.

**friable**   Material that may be easily crumbled, pulverized, or reduced to powder.

**friction factor**   A measure of the resistance to liquid flow that results from the wall roughness of a pipe or channel.

*FrictionFlex®*   Containment liner texturing process by SLT North America, Inc.

*Fridgevap®*   Package distillation/concentration system by Licon, Inc.

*Frontloader*   Reciprocating rake bar screen by Schreiber Corp.

*Frontrunner*   Reciprocating rake bar screen by Jones and Attwood, Inc.

**frothing**   The formation of a thick layer of froth on an aeration basin.

**FRP**   Fiberglass reinforced plastic.

**fry**   Juvenile fish.

**FTIR**   See "Fourier transform infrared spectrometry."

**FTU**   See "formazin turbidity unit."

*Fuchs ATAD*   Autothermal aerobic digestion system by Krüger, Inc.

**fugitive emission**   Air pollutants emitted to the atmosphere other than those from chimneys, stacks, or vents.

**fugitive source**   Any source of emissions not controlled by an air pollution control device.

**fuller's earth**   A fine, clay-like substance.

*Full-Fit™*   Membrane separator by Osmonics, Inc.

**fulvic acid**   Byproduct of decomposing organic matter that colors water.

**fume**   Finely divided solids trapped in vapor in a gas stream.

**fungi**   Small, multicellular nonphotosynthetic organisms that feed on organic matter.

**fungicide**   A substance used to kill, or inhibit the growth of, fungi or molds.

**furans**   A family of toxic chlorinated organic compounds present in minute amounts in the air emissions from hazardous waste incinerators.

**fusion**   An energy-producing nuclear reaction that results from combining nuclei of small atoms to form larger atoms.

*Futura-Thane*®   Potable water tank linings by Futura Coatings.

*Fuzzy Filter*®   Upflow filter system by Schreiber Corp.

**fuzzy logic**   A process control system intended to replace a skilled human operator by using multilevel logic to adjust process operation based on a set of approximate, rather than exact, rules.

**FWKO**   See "free water knockout."

**FWPCA**   See "Federal Water Pollution Control Act."

# G

**G**  See "velocity gradient."

**gabion**  A wire mesh container filled with rocks; used to prevent soil erosion.

**GAC**  Granular activated carbon.

**gain output ratio (GOR)**  A measure of evaporator performance that represents the ratio of mass flow of distillate to steam input.

**gallon (U.S.)**  A unit of volume for liquid substances equal to 231 cubic inches and approximately equivalent to 3.785 liters.

**gallon, imperial**  A unit of volume for liquid substances equal to 4.546 liters or 1.2 U.S. gallons.

**galvanic corrosion**  Corrosion that occurs when two dissimilar metals are connected electrically and immersed in a conductive liquid.

**galvanic series**  The ranking of the relative nobility of different conducting materials in a certain environment.

**galvanize**  An electrolytic or hot dipping process to coat steel products with zinc to increase corrosion resistance.

**gamma ray**  A short-wavelength, high-energy form of electromagnetic radiation.

**garbage**  Solid wastes generated by the handling, storage, preparation, cooking, and serving of food.

***Gard***  Trickling filter rotary distributor by General Filter Co.

***Gar-Dur***  Ultra-high-molecular-weight plastic products for chain and flight sludge collectors by Garland Manufacturing Co.

**garnet**   A dense mineral often used as media in a granular media filter.

**gas**   One of the three states of matter having no fixed shape or volume, and capable of expanding indefinitely.

**gas chromatography**   An analytical technique used to determine and measure the volatile materials in a water sample.

**gaseous emission**   Volatile or uncondensed compounds discharged into the atmosphere.

**gasification**   See "coal gasification."

*GasLifter*   Anaerobic digester circulation and mixing system by Walker Process Equipment Co.

**gasohol**   A motor vehicle fuel containing 80–90% unleaded gasoline and 10–20% ethanol.

**gas tight**   Operating with no detectable emissions.

**gastroenteritis**   An inflammation of the stomach and intestinal tract.

**gastrointestinal**   Related to the stomach or intestines.

**gate valve**   A valve with a disk-shaped closing element that slides over the opening through which water flows.

**GBT**   See "gravity belt thickener."

**GC**   See "gas chromatogragh."

**GC/MS**   Gas chromatography/mass spectrometry.

*GDT Process*™   Gas-degas treatment process for volatile organic compound removal by Mazzei Injector Corp.

**gear pump**   A positive displacement pump where cavities created between the gear teeth of two meshing gears move from the suction to the discharge side of the pump.

*GEHO®*   Piston-type pump for heavy sludges by Envirotech Company.

**Geiger counter**   An instrument used to detect radiation.

*Gelex*   Standards used to standardize turbidimeters by Hach Company.

*Gemco*   Spent liquor filter by Gauld Equipment Sales Co.

*Gemini*   Self-cleaning basket strainer by S.P. Kinney Engineers, Inc.

*Gemini*   Granular activated carbon contactors by Roberts Filter Manufacturing Co.

*Gemini Polymaster*   Emulsion and solution polymer blending equipment by Komax Systems, Inc.

*Gen2®*   Chemical feed system by Stranco, Inc.

**generator**   (1) Any person, group, or organization whose activities produce hazardous waste. (2) A rotating device used to produce electrical power.

*Generox™*   Chlorine dioxide generator by Ashland Chemical, Drew Division.

*Genesis*   Package gravity filter system by Roberts Filter Manufacturing Co.

*Genesis*   Sewage shredder by International Shredder, Inc.

*Gen-Ozi*   Ozone generator by Matheson Gas Products.

*Geoguard®*   Groundwater sampling product by American Sigma.

*Geothane®*   Wastewater containment lining by Futura Coatings, Inc.

**geothermal**   Energy produced by the transfer of heat from the earth's interior and conducted to the surface by hot water or steam.

**germ**   A disease-producing microbe.

*Gewe*   Inclined-plate settler by Purac Engineering, Inc.

*GFC*   General Filter Company.

*GFS*   Gravity Flow Systems, Inc.

**ghanat**   See "falaj."

*Giardia lamblia*   A protozoan parasite responsible for giardiasis.

**giardiasis**   Gastrointestinal disease caused by the ingestion of waterborne *Giardia lamblia*, often resulting from the activity of beavers, muskrats, or other warm-blooded animals in surface water used as a potable water source.

*Girasieve®*   Externally fed rotating drum screen by Andritz Sprout-Bauer S.A.

*Gladiator*   Groundwater remediation pump by Ejector Systems, Inc.

*GLASdek*   Synthetic media for evaporative cooling systems by Munters.

**GLI**   See "Great Lakes Initiative."

**global warming**   A theory that predicts the warming of the atmosphere as a result of an accumulation of atmospheric carbon dioxide.

**globe valve**   A valve where closure is accomplished by a horizontal plug that is lowered onto a matching seat in the center of the valve.

*Glydaseal*   Sluice gate by Rodney Hunt Co.

*G/O*   Coarse-bubble diffuser by G-H Systems, Inc.

*Golfwater®*   Aeration system for golf course ponds by Aeration Industries, Inc.

**Gooch crucible**   A heat-resistant container fitted with a filter mat used for determination of suspended and total solids.

**GOR**   See "gain output ratio."

*Gore-Tex®*   Microporous membrane material by W.L. Gore & Associates, Inc.

**gpd**   Gallons per day.

**gpg**   See "grains per gallon."

**gpg imp**   Grains per imperial gallon.

**gpm**   Gallons per minute.

*Grabber*   Reciprocating rake bar screen by Hycor, Inc.

*Grabber*   Tramp metal and heavy object catcher by Franklin Miller.

**grab sample**   A single water or wastewater sample taken at a time and place representative of total discharge.

**grade** (1) The finished surface of a civil structure. (2) The inclination or slope of a surface or structure. (3) To rate according to a standard or size.

**gradient** The rate of change of an elevation, velocity, or pressure per unit length.

**graduated cylinder** A glass cylinder with fine gradations; used in a laboratory to measure liquid contents.

**grains per gallon (gpg)** A unit of measure where 1 gpg = 17.1 mg/L; frequently used in water hardness calculations.

**gram** A unit of mass equal to the weight of 1 cm³ (1 mL) of water, or 0.03527 oz.

**gram stain** A staining procedure used to differentiate and categorize bacteria.

**granular activated carbon (GAC)** A granular form of activated carbon, used in filter beds or contactor vessels to absorb organic compounds.

**granular media filtration** A tank or vessel filled with sand or other granular media to remove suspended solids and colloids from a water or wastewater that flows through it.

*Gravabelt* Gravity belt thickener by Komline-Sanderson.

**gravel** Rock fragments measuring 2–70 mm; often used as support material in granular media filters.

*Gravex* Zeolite softener by Graver Co.

*Gravilectric* Sludge wasting system using load cells to determine excess sludge accumulations by Patterson Candy International, Ltd.

*Gravi-Merik*™ Gravimetric belt feeder by Merrick Industries.

**gravimetric** Pertaining to the measurement of the weight of samples or materials.

**gravimetric feeder** Dry chemical feeder that supplies a constant weight of chemical over a preset time period.

*Gravipak* Crossflow inclined-plate clarifier by Aerators, Inc.

*Gravi-Pak* Oil/water separator by the former Bowser-Briggs Filtration Co.

*Gravisand*™ Traveling bridge filter components by Davis Water & Waste Industries, Inc.

**gravitational acceleration** The acceleration of a free-falling body in a vacuum equal to 9.8 m (32 ft) per second per second.

*Gravitator*™ Clarifier/thickener by DAS International, Inc.

**gravity belt thickener (GBT)** A sludge dewatering device utilizing a porous filter belt to promote gravity drainage of water.

**gravity filter** Granular media filter that operates at atmospheric pressure.

**gravity spring** See "seepage spring."

**gravity system** A hydraulic system that relies on gravity flow and does not require pumping.

**gravity thickening** A sedimentation basin designed to operate at high solids loading rates, usually with vertical pickets mounted to the revolving sludge scrapers to assist in releasing entrained water.

**gray water** All nontoilet household water including the water from sinks, baths, and showers. Also called "sullage."

**grease** Common term used for fats, oils, waxes, and related constituents found in wastewater.

*GreaseBurn* Grease and skimmings incinerator by Walker Process Equipment Co.

*Grease Grabber* Grease and oil skimmer by Abanaki Corp.

**grease trap**   A receptacle used to collect grease and separate it from a wastewater flow.

**Great Lakes Initiative (GLI)**   Proposed guidelines to develop uniform water quality requirements for the Great Lakes, U.S. basin.

**green belt**   An undeveloped area immediately surrounding a town or development for the purpose of restricting indiscriminate outward expansion.

**greenhouse effect**   The effect of $CO_2$ and other gases on the earth's atmosphere that is analogous to greenhouse glass because it restricts the outflow of radiative energy, which results in the warming of the lower atmosphere.

**greenhouse gases**   Gases including carbon dioxide, methane, nitrous oxide, and chlorofluorocarbons that have been recognized to contribute to the greenhouse effect.

*Greenleaf Filter Control*   Multiple-cell rapid sand gravity filter using a central control and backwashing system by Infilco Degremont, Inc.

**green liquor**   The liquor resulting from dissolving molten smelt from the kraft recovery furnace in water.

**greensand**   A filter sand containing glauconite with ion exchange properties; often used to remove iron or manganese from water.

*Griductor®*   Comminutor by Infilco Degremont, Inc.

*Griffin Generator*   Ozone generator by Ozonia North America.

*Grifter®*   Packaged pumping/grinding station by Ingersoll-Dresser Pump (U.S.) and H2O Waste-Tec (U.K.).

*Grind Hog™*   Mechanical shredding device by G.E.T. Industries, Inc.

**grit**   Sand, gravel, cinders, and other heavy solid matter that have settling velocities substantially higher than those of organic putrescible solids in wastewater.

**grit chamber** A settling chamber used to remove grit from organic solids through sedimentation or an air-induced spiral agitation.

**grit classifier** Mechanical device utilizing an inclined screw or reciprocating rake to wash putrescible organics from grit.

*Grit King*™ Grit removal unit by H.I.L. Technology, Inc.

*GritLift* Airlift pump for grit removal by Walker Process Equipment Co.

*Gritmeister*™ Grit separator/screw conveyor by Hycor Corp.

*Gritreat* Aerated grit chamber by Envirex, Inc.

**grit removal** A preliminary wastewater treatment process to remove grit from organic solids.

*Grit Snail*™ Fine grit removal system by Eutek Systems, Inc.

**grit washer** A device used to wash organic matter from grit.

**groundwater** Subsurface water supply contained in underground aquifers or reservoirs.

**grout** Fluid, or semifluid, cement slurry for pouring into joints of brickwork or masonry.

**GRP** Glassfiber reinforced plastic.

**grubbing** The process of removing tree stumps or roots.

*Guardian*™ Strainer product line by Tate Andale, Inc.

**guide vane** A device used to direct or guide the flow of a liquid or vapor.

*Gundline*® Containment liner by Gundle Lining Systems, Inc.

*Gundnet* Drainage net by Gundle Lining Systems, Inc.

*Gundseal* Geocomposite liner by Gundle Lining Systems, Inc.

**G value** See "velocity gradient."

**GWDR** Ground Water Disinfection Rule.

**GWUI**  Groundwater under the direct influence of surface water.

**gypsum**  Mineral consisting primarily of fully hydrated calcium sulfate. Chemical formula is $CaSO_4 \cdot 2H_2O$.

*Gyrazur*™  Softening clarifier by Infilco Degremont, Inc.

# H

**ha**   See "hectare."

**HAA**   Haloacetic acids.

**HAA5**   Five haloacetic acids which include mono-, di-, and trichloroacetic acids, and mono- and dibromoacetic acids.

*Hach One*   pH electrode by Hach Co.

*Halberg*   Digester draft tube sludge mixer by SIHI Pumps, Inc.

**half-life**   The time required for half of the atoms of a particular radioactive substance to transform, or decay, to another nuclear form.

**half-life, chemical**   The time required for the concentration of a chemical being tested to be reduced to one-half of its initial value.

**halide**   A compound containing a halogen.

**halocline**   A well-defined salinity gradient in an ocean.

**halogen**   One of the chemical elements of the group containing fluorine, chlorine, bromine, iodine, and astatine.

**halogenated organic compounds (HOC)**   Compounds having a halogen-carbon bond.

**halophyte**   Plants capable of living in salt water or salty soil.

*Hamburg Rotor*   Surface aerator by Hellmut Geiger GmbH & Co.

*Hammer-Head*™   Groundwater pump by QED Environmental Systems.

**hammermill**   A device with hammer-like arms used to shred or grind solids to facilitate further treatment or disposal.

*Hankin*™   Ozone systems product line by Wheelabrator Clean Water Systems, Microfloc Products.

**HAP**   Hazardous air pollutant.

*Harbor Bosun*   Dye tablets by Formulabs, Inc.

*Harborlite*®   Perlite material by Celite Corp.

*HaRDE*®   Electrostatic precipitator by Wheelabrator Clean Water Systems, Inc.

*Hardinge*   Manufacturer of traveling bridge filter whose product line was acquired by Infilco Degremont, Inc.

**hardness**   The total concentration of calcium and magnesium ions in water expressed as calcium carbonate.

*Hastelloy*®   Corrosion-resistant, nickel-based alloy by Haynes International, Inc.

*Hawco*   Former name of Screening Systems International.

*Hawker Siddeley Brackett*   Former parent company of Brackett Green, Ltd.

**Hazardous and Solid Waste Amendments (HSWA)** 1984 amendments to the Resource Conservation and Recovery Act, regulating underground tank storage and land disposal of certain hazardous wastes.

**hazardous area, class 1**   Locations where flammable gases or vapors may be present in the air in sufficient quantities to produce explosive or ignitable mixtures.

**Hazardous Ranking System (HRS)**   The method used to evaluate the relative potential of hazardous substance releases to cause health or safety problems, or ecological or environmental damage.

**hazardous waste**   Any waste or combination of wastes that pose a substantial present or potential hazard to human health or living organisms, because they are nondegradable, persistent in nature, or may otherwise cause detrimental cumulative effects.

**Hazen-Williams coefficient** A roughness coefficient related to the influence of pipe material on the velocity characteristics of a fluid.

**hazmat** Hazardous materials.

**HAZOP** Hazard and Operability Study.

**HAZWOPER** Hazardous Waste Operation and Emergency Response.

**HCFC** See "hydrochlorofluorocarbon."

**HDPE** See "high-density polyethylene."

**head** (1) A measure of the pressure exerted by a fluid expressed as the height of an enclosed column of the fluid that could be balanced by the pressure in the system. (2) The source or upper end of a system, e.g., headwater.

**header** A pipe manifold fitted with several smaller lateral outlet pipes.

**headloss** The difference in water level between the upstream and downstream sides of a treatment process.

**headwater** The source or upper reaches of a stream.

**headworks** The initial structure and devices located at the receiving end of a water or wastewater treatment plant.

**hearth** The bottom of a furnace, upon which waste materials are exposed to the flame.

*Heatamix* Heating and recirculation system for anaerobic sludge digesters by Simon-Hartley, Ltd.

**heat balance** An accounting of the distribution of a system's heat loss and heat gain.

**heater treater** Oil field produced water treatment unit used to break water-in-oil emulsions with the heat and chemicals.

**heat exchanger** A device used to transfer heat from one substance to another. See also "shell-and-tube heat exchanger."

**heat island**   An urban phenomenon where air pollutants and the heat from a combination of tall buildings, concrete pavement, and other materials combine to create a localized haze dome that traps rising hot air, resulting in higher temperatures and air pollution.

**heat of condensation**   The amount of heat released when a vapor changes state to a liquid.

**heat of vaporization**   The amount of heat required to change a volume of liquid to a vapor.

**heat pump**   A device for transferring heat from a cooler reservoir to a hotter one by mechanical means, involving the compression and expansion of a fluid.

**heat sink**   Any material that is used to absorb heat.

*Heat Systems*   Former name of Misonix, Inc.

**heat tracing**   The electrical or steam heating of piping and equipment to prevent freezing.

**heat transfer**   The transfer of heat from one body to another by means of radiation, conductance, or convection.

**heat value**   The quantity of heat that can be released from a sludge per unit mass of the sludge solids.

*HeatX*   Anaerobic digester gas heating unit by Walker Process Equipment Co.

**heavy metals**   Metals that can be precipitated by hydrogen sulfide in an acid solution, and which may be toxic to humans above certain concentrations.

**hectare (ha)**   A unit of area equal to 10,000 square meters. One hectare is equivalent to 2.471 acres.

**HEEB**   High-energy electron beam irradiation.

*Hela-Flow*   Plastic laterals for water and wastewater treatment by Liquid-Solids Separation Corp.

*Helaskim*   Helical surface skimmer by Walker Process Equipment Co.

*HeliCarb*   Carbon dioxide contactor by CBI Walker.

*Heliclean*® Open-channel spiral screen with screenings washer by Hycor Corp.

*Helico* Screw-type screenings press by Infilco Degremont, Inc.

*Helicobacter pylori* A bacterium that causes stomach ulcers and has been identified as an emerging water-borne health threat.

*Heli-Press* Screenings compactor by Vulcan Industries, Inc.

*Helisieve*® Open-channel spiral screen by Hycor Corp.

*Helisieve Plus*™ Septage receiving station by Hycor Corp.

*HeliSkim* Helical surface skimmer by Walker Process Equipment Co.

*HeliThickener* Interrupted flight screw conveyor for sludge collectors by Walker Process Equipment Co.

*Helixor* Subsurface aerator/mixer by Polcon Sales Ltd. (North America) and Environmental Construction Ltd. (U.K.).

*Helixpress* Screenings dewatering press/conveyor by Hycor Corp.

helminth A parasitic worm.

hemi-hydrate A crystalline compound having one molecule of water of crystallization per two molecules of compound.

Henry's law The weight of any gas that will dissolve in a given volume of a liquid, at constant temperature, is directly proportional to the pressure that the gas exerts above the liquid.

HEPA See "high-efficiency particulate air filter."

hepatitis An acute viral disease that results in liver inflammation and may be transmitted by direct contamination of a water supply by sewage.

**herbicide**   A synthetic organic compound used to control plant growth.

*Hercules*   Pressure leaf filters by Liquid-Solids Separation Corp.

*Hercules*   Screening equipment division of Atlas Polar Co.

**heterotrophic bacteria**   Bacteria that derives its cell carbon from organic carbon; most pathogenic bacteria are heterotrophic bacteria.

**heterotrophic plate count (HPC)**   A laboratory method of determining the level of heterotrophic bacteria in a sample.

*Hevi-Duty*   Traveling water screen replacement parts products by Envirex, Inc.

*Hex-Air*   Fine-bubble air diffuser system by Dunlop, Ltd.

**hexavalent chrome**   A toxic form of chrome used in plating operations, usually reduced to the trivalent form and precipitated as a hydroxide.

**HFC**   See "hydrofluorocarbon."

**HHW**   Household hazardous waste.

*Hi-Capacity*   Clarifier/thickener by Eimco Process Equipment Co.

*Hidrostal*   Screw/centrifugal impeller pump by Envirotech Company.

*Hi-Flo®*   Depth filter and water softener products by Culligan International Corp.

**high-density polyethylene (HDPE)**   A synthetic organic material often used as landfill liner because of its low permeability.

**high-efficiency particulate air filter (HEPA)**   A filtering system capable of trapping and retaining at least 99.97% of all monodispersed particles 0.3 μm in diameter or larger.

*High Flux Series*™   Membrane cleaning products for restoring membrane flux by King Lee Technologies.

*High-Flux TF™*   Oil control chemical for use in a membrane treatment system by King Lee Technologies.

**high-level radioactive waste**   The highly radioactive waste material resulting from the reprocessing of spent fuel, the spent fuel itself, and much of the waste generated from nuclear weapons production, with activities measured in curies per liter.

*High-Line Combi*   Rotating drum screen by Hercules Systems Ltd.

**high-performance liquid chromatography (HPLC)**   Instrumental technique for measuring trace levels of organics by means of UV adsorption.

**high-test calcium hypochlorite**   Calcium hypochlorite product containing at least 70% available chlorine.

**high-velocity air filter (HVAF)**   An air pollution control filtration device for the removal of sticky, oily, or liquid aerosol particulate matter from exhaust gas streams.

*Hi-Iron*   Closed-pressure, contact-aeration iron removal process by Aquatrol Ferr-X Corp.

*Hi-Lift*   Reciprocating rake bar screen by Longwood Engineering Co.

*Hi-Lucid*   High-rate coagulation and sedimentation process by Hitachi Metals America, Ltd.

**hindered settling**   Sedimentation of particles in a suspension of intermediate concentration where interparticle forces hinder the settling of neighboring particles.

*HIP*   High-intensity belt press by Andritz-Ruthner, Inc.

*Hi-Pass*   In-line motionless mixer by Komax Systems, Inc.

*Hi-Rate Thickener*   Circular gravity sludge thickening system with ancillary equipment by Dorr-Oliver, Inc.

*Hi-Tork*   Portable mixer by Philadelphia Mixer Corp.

**HMIP**   Her Majesty's Inspectorate of Pollution (U.K.).

**H₂O**   Chemical formula for water.

**H₂O₂**   See "hydrogen peroxide."

**HOA**   Hand-off-automatic.

**HOC**   See "halogenated organic compounds."

*HOD-WACX*   Clarifier incorporating buoyant filtration media by Dean Wacx.

*H₂O Express*   Belt dewatering press by Magnet Machinery, Inc.

*HOG™*   Halogenated organic gas destruction process by Quantum Technologies, Inc.

**hogging ejector**   A steam jet ejector that produces a vacuum in a vessel by evacuating the contents, in a single stage, to the atmosphere.

**holding tank**   A tank used to receive and store wastewater prior to its ultimate disposal.

*Hollosep*   Hollow fiber configuration reverse osmosis module by Toyobo Co., Ltd.

**hollow fiber**   Type of reverse osmosis and ultrafiltration membranes formed into small-diameter tubes.

*Homomix*   Propeller-type rapid mixer by Amwell, Inc.

**HON**   Hazardous Organic NESHAP (National Emission Standards for Hazardous Air Pollutants). A rule regulating emissions of listed organic chemicals from new and existing manufacturing sources.

**honey wagon**   Common term for vacuum truck used to remove accumulated septic tank sludge.

*Hosepump*   Peristaltic pump by Waukesha Pumps, Inc.

**host**   Any plant or animal on or in which another lives for nourishment, development, or protection.

**hot lime softening**   Lime softening process conducted at temperatures of 104–125°C.

**hp**   Horsepower. A standard unit of power equal to 745.7 watts.

**HPC**   See "heterotrophic plate count."

*HPD*   Evaporator product group of Wheelabrator Engineered Systems, Inc., HPD Division.

*HP-Hybrid*   Automatic filter press by Heinkel Filtering Systems, Inc.

**HPLC**   See "high-performance liquid chromatography."

*H. pylori*   See *"Helicobacter pylori."*

*HRB®*   Solid waste baler by Harris Waste Management Group, Inc.

**HRGC**   High-resolution gas chromatography.

**HRMS**   High-resolution mass spectrometry.

**HRS**   See "Hazardous Ranking System."

**HRSG**   Heat recovery steam generator.

**HRT**   See "hydraulic residence time."

**H$_2$S**   See "hydrogen sulfide."

*HSC*™   High-pressure centrifugal pump for reverse osmosis by Pump Engineering, Inc.

**HSWA**   See "Hazardous and Solid Waste Amendments."

**HTA**   High-temperature additive.

**HTC**   Heat transfer coefficient.

*HTC*™   Hydraulic turbocharger for reverse osmosis pump system by Union Pump Company.

*HTH*   High-test calcium hypochlorite product by Olin Corp., Chemicals Group.

**HTP**   Heat-treated peat, polymer spill encapsulation, and cleanup product by American Products.

*Huisman*   Oxidation ditch wastewater treatment system by Envirex, Inc.

**humic acid**   Organic acids that are byproducts of decomposing organic matter that colors water.

**humidity**   The amount of water vapor within the atmosphere.

**humus**   Dark or black decomposing organic matter in soil.

*Hurricane*   Mixer by Franklin Miller.

*Hurricane*   Combination centrifuge and cartridge filtration system by Harmsco Industrial Filters.

**HVAC**   Heating, ventilating, and air conditioning.

**HVAC/R**   Heating, ventilating, air conditioning, and refrigeration.

**HVAF**   See "high-velocity air filter."

**HWIR**   Hazardous Waste Identification Rule.

**HWL**   High water level.

**HX**   Heat exchanger.

**hyacinth**   Floating aquatic plants the roots of which provide a habitat for a diverse culture of aquatic organisms that metabolize organics in wastewater.

**hybrid system**   A system incorporating multiple processes or technologies.

*Hydecat™*   Hypochlorite destruction product by ICI Katalco.

*HYDRA*   Hydraulic rake sludge removal system by Hazleton Environmental Products, Inc.

*Hydradenser*   Inclined screw thickener by Black Clawson, Shartles Division.

*Hydra-Mix®*   Hydraulic mixer by Air-O-Lator Corp.

*Hydrapaint*   Ultrafiltration spiral membrane by Hydranautics.

*Hydra-Press™*   Hydraulic screenings compactor by Vulcan Industries.

*Hydrasand®*   Continuous cleaned moving bed sand filter by Andritz-Ruthner, Inc.

*Hydrasieve®*   Static fine screen by Andritz-Ruthner, Inc. (Western Hemisphere) and Andritz Sprout-Bauer S.A. (Eastern Hemisphere).

**hydrate**   A compound formed by the union of water with another substance.

**hydrated lime**   The product that results from mixing quicklime with water. Chemical formula is $CaOH_2$.

**hydration** The process of combining or uniting water with another substance.

*Hydra-Tracker*™ Self-tracking oil, grease, and sludge removal device by Dontech, Inc.

*Hydraucone* Diffuser plate used with down flow mechanical aerator by Amwell, Inc.

**hydraulic gradient** The slope of the hydraulic grade line, which indicates the change in pressure head per unit of distance.

**hydraulic jump** The sudden rise in water surface level that may occur when water flowing through an open channel at a high velocity is retarded.

**hydraulic loading** Total volume of liquid applied per unit of time to a tank or treatment process.

**hydraulic radius** The ratio of the area of a conduit in which water is flowing to its wetted perimeter.

**hydraulic residence time (HRT)** Vessel volume divided by the liquid removed.

**hydraulics** The branch of science and engineering that deals with the mechanics of fluids.

**hydrazine** A chemical compound used as an oxygen scavenger. Chemical formula is $H_2N_4$.

**hydroblast** The use of high-pressure water jets to clean or remove debris.

*Hydroburst* Passive screen air backwash system by Wheelabrator Engineered Systems, Inc., Johnson Screens.

**hydrocarbons** Organic compounds consisting predominantly of carbon and hydrogen.

*HydroCeal* Mixed-bubble diffuser by Hydro-Aerobics, Inc.

*Hydrocell*® Induced-air flotation separators by U.S. Filter Corp.

*Hydro-Chek*   Coarse-bubble diffuser by Pollution Control, Inc.

**hydrochloric acid**   An aqueous form of hydrogen chloride that is a strong corrosive agent. Chemical formula is HCl.

**hydrochlorofluorocarbon (HCFC)**   A substance used as a temporary alternative to a CFC.

*Hydro-Circ®*   Nonmechanical sludge recirculation system by Graver Co.

*Hydrocleaner*   Dissolved-air flotation aeration system by Envirotech Company.

*Hydro Clear®*   Pulsed bed gravity sand filter by Zimpro Environmental, Inc.

*HydroClor-Q™*   Organic chlorine test kit by Dexsil Corp.

*Hydro-Cone*   Underdrain for sand filter by Leeds & Northrup, a unit of General Signal.

**hydrocyclone**   A conical-shaped device that utilizes centrifugal force to separate grit and other solids from a liquid.

*Hydrodarco®*   Activated carbon by Norit Americas, Inc.

*HydroDri™*   Screenings press by Serpentix Conveyor Corp.

*HydroFlo™*   Disposable in-line filter for groundwater monitoring by Schlicher & Schuell.

*Hydro-Float*   Flotation system for removal of fat, grease, and suspended solids by HydroCal Company.

*Hydrofloc*   Polyelectrolyte used to enhance liquid/solid separation by Aqua Ben Corp.

*HydroFloc™*   Rotating screen thickener by Klein America, Inc.

*Hydrofluor Combo*   Cryptosporidium and Giardia detection reagents by Meridian Diagnostics, Inc.

**hydrofluoric acid**   An aqueous form of hydrogen fluoride. Chemical formula is HF.

**hydrofluorocarbon (HFC)**   A replacement for chlorofluorocarbons and hydrochlorofluorocarbons that contains no chlorine and has no ozone-depletion potential.

*Hydroflush*   Cable-operated bar screen by Beaudrey Corp.

**hydrogen peroxide**   An oxidizing agent used for odor control and disinfection. Chemical formula is $H_2O_2$.

**hydrogen sulfide**   A toxic gas formed by the decomposition of organic matter containing sulfur. Chemical formula is $H_2S$.

**hydrograph**   A graphical representation of a stream discharge at a single location.

*Hydrogritter*   Grit washing system by Envirotech Company.

*Hydro-Lift*   Prefabricated steel lift stations by Hydro-Aerobics, Inc.

*Hydro-lite*   Biological filter media by Hydro-Aerobics, Inc.

**hydrologic cycle**   The natural cycle of continuous evaporation and condensation.

**hydrolysis**   A chemical change or decomposition of matter produced by combination with water.

*Hydromaster*   Packaged water treatment plant by Gladwall Engineering Services, Ltd.

*Hydromation*   Former equipment manufacturer acquired by Filtra-Systems.

**hydronium ion**   The hydrated hydrogen ion, $H_3O^+$.

*Hydroperm®*   Crossflow microfiltration system by Wheelabrator Engineered Systems, Inc., Microfloc Products.

**hydrophilic**   Having an affinity for water.

**hydrophobic**   Having an aversion to water.

**hydrophyte**   A plant that grows in water or saturated soils.

*Hydropillar™*   Elevated water storage tank by Pitt-Des Moines, Inc.

*Hydropress*   Belt filter press by the former Clow Corp.

*Hydro-Press*   Belt filter press by HydroCal Company.

*HydroPunch®*   Groundwater sampler by QED Environmental Systems.

*Hydrorake*   Trash rake by Hercules Division/Atlas Polar Co.

*HydroRanger*   Ultrasonic level measuring system by Milltronics.

*Hydro-Rotor*   Brush-type aerator by Amwell, Inc.

*Hydro-SAFe*   Biological aerated filter by Hydro-Aerobics, Inc.

*Hydro Scour*   Pulsed bed filter backwashing system by Zimpro Environmental, Inc.

*Hydroscreen*   Static fine screen by Hycor Corp.

*HydroSeal®*   Anaerobic digestion gasholder cover by Eimco Process Equipment.

*Hydrosep*   Shallow basin, nonaerated grit removal system by Aerators, Inc.

*Hydroseparator*   Industrial wastewater gravity thickener by Dorr-Oliver, Inc.

*Hydroshear*   Aeration tank for low-flow package treatment plants by FMC Corp., MHS Division.

*Hydro-Shear*   Internally fed rotary fine screen by Dontech, Inc.

*Hydrosil*   Static screen with brush/water jet cleaning device by Spirac AB (Europe) and JDV Equipment Corp. (U.S.).

*Hydro-Sock*   Upflow cartridge filtration system by Hydro-Aerobics, Inc.

**hydrosphere**   The aqueous environment of the earth including rivers, lakes, oceans, and glaciers.

**hydrostatic pressure**   The pressure exerted by water due to depth alone.

**hydrotest**   The testing of piping, tubing, or vessels by filling with water and pressurizing to test for integrity.

*Hydrovex™*  Flow control product line by John Meunier, Inc.

*Hydrowash*  Downflow recycling pump/aerator for grit removal by Amwell, Inc.

**hydroxide**  A negatively charged ion consisting of a hydrogen atom and an oxygen atom. Chemical formula is OH.

**hydroxide alkalinity**  Alkalinity caused by hydroxyl ions.

**hydroxyl**  A chemical group consisting of one hydrogen and one oxygen atom.

*Hydro-Zap®*  Ultraviolet wastewater disinfection systems by Hydro-Aerobics, Inc.

*Hydrozon®*  Ozone system by Carus Chemical Co.

*Hydrymax*  Sludge dryer by D.R. Sperry & Co.

**hyetograph**  (1) A graphic representation of a rainfall that plots time versus rainfall. (2) A rain recording gauge.

*Hy Flo Super-Cel®*  Diatomaceous earth filter media by Celite Corp.

*Hygene*  Bacteriostatic filter media by Ionics, Inc.

**hygrometer**  An instrument for measuring the relative amount of moisture in the air.

**hygroscopic**  Readily absorbing moisture from the atmosphere.

**hygroscopic water**  Water in soil that is in equilibrium with the atmospheric water vapor; it cannot be lost to evaporation or drain freely by gravity.

*Hymergible®*  Hydraulically driven submersible pump by Crane Pumps & Systems.

**hyperfiltration**  See "reverse osmosis."

*HyperFlex®*  High-density polyethylene containment liner by SLT North America, Inc.

*Hyperfloc®*  Polyelectrolyte used to enhance liquid/solid separation by Hychem, Inc.

*Hyper+Ion*™   Cationic coagulants and flocculants by General Chemical Corp.

*HyperNet*™   Net fabric composite by SLT North America, Inc.

*Hyperpress*™   Combination belt filter/plate and frame press by Klein.

*Hypersperse*   Antiscalant for use in reverse osmosis systems by Argo Scientific.

**hypochlorite**   Chlorine anion commonly used as an alternative to chlorine gas for disinfection. Chemical formula is $OCl_3^-$.

*Hypo-Gen*®   Sodium hypochlorite generation systems by Capital Controls Co., Inc.

**hypolimnetic aeration**   Aeration of water at the bottom of a lake.

**hypolimnion**   The lower layer of a stratified lake that results from varying water densities.

**hypoxia**   The condition that exists when a water has a very low dissolved water level, usually less than 2 mg/L.

*Hypress*   Screenings dewatering press by Hycor Corp.

*Hy-Q*   Flush bottom sluice gate closure by Rodney Hunt Co.

*Hysep*®   Decanter centrifuge by Centrico, Inc.

*Hy-Speed*®   Mixer by Alsop Engineering Co.

*Hytrex*®   Cartridge prefilters and prefilter housings by Osmonics, Inc.

*Hyveyor*   Troughing conveyor by Hycor Corp.

# I

**I/A**   Innovative and alternative.

**IAF**   See "induced-air flotation."

**IAQ**   Indoor air quality.

**IBWA**   International Bottled Water Association.

**I&C**   Instrumentation and control.

**ICE**   Institute of Civil Engineers.

*ICEAS®*   Intermittent cycle extended aeration system for wastewater treatment by Austgen Biojet Wastewater Systems.

*Ice-Away®*   Ice melter by Air-O-Lator Corp.

**ice fog**   An atmospheric suspension of reflective ice crystals that affects visibility.

**ID**   Inside diameter.

**IDA**   International Desalination Association.

*IDI®*   Infilco Degremont, Inc.

**IDLH**   See "immediately dangerous to life and health."

*IDP*   Ingersoll Dresser Pump Co.

*Idrex*   Pressure leaf filter by Zimpro Environmental, Inc.

*IDS Drumshear*   Rotating fine screen by Aer-O-Flo Environmental, Inc.

**IEC**   International Electrotechnical Commission.

**IEEE**   Institute of Electrical & Electronic Engineers.

*IFU*   Fine-bubble membrane diffuser by Envirex, Inc.

**IGCC**   Integrated coal gasification combined cycle.

**igneous rock**   A type of rock formed from cooled magma.

**ignitability**   The characteristic of having a flash point less than 60°C.

**I/I**   See "inflow/infiltration."

**IIA**   Incinerator Institute of America.

*II-PLP®*   Double-pass reverse osmosis system with inter-stage chemical feeds for pH adjustment by U.S. Filter Corp.

**ilmenite**   A dense mineral often used as filter media in a granular media filter.

**imbibition**   The process of assimilating or taking into solution.

**Imhoff cone**   Cone-shaped container used to determine settleable solids.

**Imhoff tank**   A two-story wastewater treatment tank in which sedimentation occurs in the upper compartment and anaerobic digestion occurs in the lower compartment.

**immediately dangerous to life and health (IDLH)** The maximum environmental concentration of a substance from which one could escape from a 30-min exposure without irreversible adverse health effects.

**immiscible**   Incapable of being mixed.

**immunoassay**   The identification of a substance based on its capacity to act as an antigen.

*Impac™*   Packing media for air stripping towers by Lantec Products, Inc.

**impact fee**   Fee assessed new connections to a water or sewer system that are intended to recover a portion of the capital cost of the system.

**impeller**   The rotating set of vanes in a turbine, blower, or centrifugal pump designed to cause rotation of a fluid mass.

**imperial gallon**   See "gallon, imperial."

**impermeable strata**   Layers of clay or dense stone in the earth through which water cannot penetrate in measurable quantities.

**impervious**   Not allowing the passage of water at ordinary hydrostatic pressure.

**impingement** (1) The entrapment of fish and other marine life on the surface of an intake screen when a high water velocity prevents escape. (2) The striking of a surface by a moving fluid.

**impoundment** A pond, lake, or reservoir created through the use of a structural barrier such as a dam, levee, or dike.

**impressed voltage cathodic protection** The use of an impressed current to prevent or reduce the rate of corrosion of a metal in an electrolyte by making the metal the cathode for the impressed current.

*Impulse®* Countercurrent water softeners by U.S. Filter Corp.

**impurity** A chemical substance that is unintentionally present with another chemical substance or mixture.

**IMS** See "Ion mobility spectrometry."

*IMS®* Filter media support cap for sand filters by F.B. Leopold Co.

**incidence of illness** The rate of occurrence of new cases of disease in a defined population over a specified period of time.

**incineration** The process of reducing the volume of a solid by burning of organic matter.

**incinerator** A furnace or device for incineration.

**inclined-plate separators** A series of parallel inclined plates that can be used to increase the efficiency of clarifiers and gravity thickeners.

**incompatible waste** A hazardous waste that may cause corrosion or decay of containment materials, or is unsuitable for commingling with another waste under uncontrolled conditions because of the hazardous reactions that may result.

**incubate** To maintain optimum environmental conditions for growth and reproduction of viable microbes.

*Incutrol*    Biochemical oxidation measuring and temperature control apparatus for biochemical oxygen demand incubation by Hach Company.

**indicator organism**    Microbes whose presence indicates the absence or presence of a specific pollutant.

**indirect discharger**    Water treatment plants that discharge pollutants to publicly owned treatment works.

**induced-air flotation (IAF)**    The clarification of suspended, dispersed material with hydrophobic surfaces by contact with dispersed air bubbles causing that attached to the hydrophobic surfaces causing the materials to collect as a froth.

**induced-draft cooling tower**    A cooling tower in which the air flow through the tower is induced by means of an electrically operated fan.

**industrial waste**    Waste generated by manufacturing or industrial practices that is not a hazardous waste regulated under Subtitle C of the Resource Conservation and Recovery Act.

**industrial wastewater**    Liquid wastes resulting from industrial practices or processes.

**inert**    Lacking active properties and unable to react with other substances.

*Infa Screen*    Multiple-rake bar screen by Biwater Treatment Ltd.

**infectious agent**    Any organism that is capable of being communicated in body tissues and capable of causing disease or adverse health impacts in humans.

**infectious waste**    Equipment, instruments, pathological specimens, or other disposable wastes that may be contaminated from persons who are suspected to have or have been diagnosed as having a communicable disease.

**infiltration**    (1) Water entering a sewer system through broken or defective sewer pipes, service connections, or manhole walls. (2) Wind-induced air movement into a building through openings in walls, doors, or windows.

**infiltration gallery**   A horizontal underground conduit of screens or porous material to collect percolating water, often under a river bed.

**inflammable**   Easily set on fire.

**inflow**   Surface and subsurface water or stormwater discharged into a sewer system.

**inflow/infiltration (I/I)**   The total quantity of water from inflow and infiltration without distinguishing the source.

**influent**   Water or wastewater flowing into a basin or treatment plant.

**infrared radiation**   Low-energy radiation with wavelengths longer than visible light and shorter than radio waves.

**infrastructure**   The fundamental network of facilities, installations, and utility systems serving a community.

**inhalable diameter**   The diameter of a particle, considered to be less than 15 μm for humans, which is capable of being inhaled and deposited anywhere within the respiratory tract.

**inhalation $LC_{50}$**   The concentration of a substance, expressed as mg/L of air, that is lethal to 50% of the test population.

**inhibitor**   A chemical that interferes with a chemical reaction.

**injection well**   A hole drilled below the ground surface into which wastewater or treated effluent is discharged.

*Innova-Tech*[sm]   Total barrier oxidation ditch wastewater system by Innova-Tech, Inc.

**innovative technology**   A process or technique that has not been fully proven under the circumstances of its contemplated use and that represents an advancement over the state of the art.

**inorganic matter**   Substances of mineral origin, not containing carbon, and not subject to decay.

**in situ**   Treatment or disposal methods that do not require movement of contaminated material.

*In-Situ Oxygenator*™   Mechanical floating aerator by Praxair, Inc.

**insoluble**   A compound that has very low solubility.

*Instant Ocean*®   Mineral concentrate used to simulate seawater salinity by Aquarium Systems.

*InstoMix*   In-line and in-channel mixers by Walker Process Equipment Co.

**intake**   The headworks or structure at the receiving end of a raw water treatment system.

*IntensAer*   Radial surface aerator formerly offered by Walker Process Equipment Co.

**interceptor sewer**   A sewer that receives flow from a number of other sewers or outlets for disposal or conveyance to a treatment plant.

**intercondenser**   A condenser used between stages to reduce steam consumption in the steam jet vacuum system in an evaporator system.

**interfacial tension**   The tension that occurs at the interface between two fluids or a liquid and a solid.

**intergranular corrosion**   Corrosion at or near the grain boundaries.

*Inter-Mix*®   Slow-speed mixer by Air-O-Lator Corp.

*Internalift*®   Enclosed screw pump by Wheelabrator Engineered Systems, Inc., CPC Engineering Products.

*International Process System*   Former supplier acquired by Wheelabrator Clean Water Systems, Inc., Bio Gro Division.

*Interox America*   Former name of Solvay Interox.

*Inter-Sep*™   Rotary screen by Dontech, Inc.

**interstice**  An open space in granular material that is not occupied by solid material.

**interstitial water**  Water contained in the interstices of rocks.

*Intracid®*  Water tracing dye by Crompton & Knowles Corp.

**inverse solubility**  The characteristic attributed to a substance that becomes less soluble with increasing temperature.

**inversion**  The abnormal atmospheric condition that occurs when the air temperature increases with elevation.

**invert**  The lowest point of the internal surface of a drain, sewer, or channel at any cross-section.

**in-vessel composting**  Sludge composting system with integral material handling and in-vessel mixing and aeration.

**in vitro**  A laboratory study conducted in glassware.

**I/O**  Input/output.

**IOC**  Inorganic chemicals.

**ion**  An electrically charged atom, molecule, or radical.

*Ionac®*  Ion exchange resins by Sybron Chemicals, Inc.

**ion exchange (IX)**  A chemical process involving the reversible exchange of ions between a liquid and a solid.

*Ion Grabber*  Electrolytic purification unit by Atlantes Chemical Systems, Inc.

**ionic strength**  A measure of solution strength based on both the concentrations and valences of the ions present.

**ionization**  The dissociation of molecules into negatively and positively charged ions.

**Ion mobility spectrometry (IMS)**  Continuous emissions analysis used to measure pollutants in gases.

**ionosphere**  The upper level of the earth's atmosphere beginning at an altitude of approximately 80 km.

*Ionpure®*  High-purity water treatment products and services by U.S. Filter Corp.

*Ion Stick®*  Electrostatic water treater for prevention of scale and fouling by York Energy Conservation.

*Iopor*  Low-pressure ultrafiltration system formerly offered by Dorr-Oliver, Inc.

**IP**  Inhalable particulates.

**IPP**  Independent power producer.

*IPS*  In-vessel composting system by Wheelabrator Clean Water Systems, Inc., Bio Gro Division.

*IQS/3™*  Programmable controller used to operate and monitor water treatment systems by Bruner Corp.

**iron bacteria**  Bacteria capable of metabolizing reduced iron. Also "crenothrix."

*Iron Remover*  Contact bed type iron removal system by Walker Process Equipment Co.

**irrigation**  The artificial application of water to meet the requirements of growing plants or grass that are not met by rainfall alone.

*ISEP®*  Continuous adsorption/desorption contactor by Advanced Separation Technologies.

**ISO**  International Organization for Standardization.

**ISO 9000**  Certification conferred to a manufacturer that has demonstrated the capability of running an integrated business from initial design through manufacture.

**ISO 14000**  Environmental management system standards for manufacturing and service industries.

**isobar**  A line on a weather map that joins all points of equal barometric pressure.

**isomer**  A chemical compound that has the same molecular formula, but different molecular structure, as another compound.

**isopleth**  A line on a map connecting points at which a certain variable has the same value.

**isotherm** A line on a weather map connecting points that have the same temperature.

**isotopes** Atoms with the same atomic number but different atomic weights.

**IU** Industrial user.

**IWEM** See "CIWEM."

*IWT®* Illinois Water Treatment product line of U.S. Filter Corp.

**IX** See "ion exchange."

# J

*J&A*  Jones and Attwood, Inc.

*Jackbolt*™  Aluminum clarifier cover by Enviroquip, Inc.

**jacking**  A method of installing pipe by forcing it into a horizontal opening with horizontal jacks.

*JackKnife*  Pivoting air header and drop pipe arrangement by Walker Process Equipment Co.

**Jackson turbidity unit (JTU)**  Unit of measure used to quantify water turbidity by observing the outline of a candle flame viewed through a water sample. Largely replaced by "NTU."

*JAC Oxyditch*  Oxidation ditch treatment system formerly offered by Chemineer, Inc.

**jar test**  A test procedure using laboratory glassware for evaluating coagulation, flocculation, and sedimentation in a series of parallel comparisons.

*Javex-12*™  Sodium hypochlorite by Javex Manufacturing Corp.

*Jayfloc*  Polyelectrolyte used to enhance liquid/solid separation by Exxon Chemical Co.

*Jeffrey*  Screening equipment product line offered by Jones and Attwood, Inc.

*Jeffrey*®  Chain, sprocket, and related component products by Jeffrey Chain Corp.

*JelClear*™  Granular filtration media with a coagulant bonded directly to the media grains by Argo Scientific.

**jet**  A stream of pressurized liquid or vapor from a nozzle or orifice.

*JETA*   Vortex-type grit collector by Jones and Attwood, Inc.

**jet aeration**   Wastewater aeration system using floor-mounted nozzle aerators that combine liquid pumping with air diffusion.

*Jeta-Matic*   Spray jet pressure leaf filter cleaning system by U.S. Filter Corp.

*Jet Breaker*™   Screenings washer/compactor by Mahr.

*Jet-Chlor*®   Tablet chlorine disinfection system by Jet, Inc.

*JETIII*®   Fabric filter by Wheelabrator Clean Water Systems, Inc.

*JetMix*   Sludge storage and mixing system by A.O. Smith Harvestore Products, Inc.

*Jet Plant*   Package wastewater treatment plant by Jet, Inc.

**jetsam**   Floating jettisoned material.

*Jet Shear*   Continuous mixing system using jet nozzles by Flo-Trend Systems, Inc.

**jet stream**   A strong, thermally driven, high-altitude wind.

*Jet-Tex*   Leach bed filter fabric by Jet, Inc.

*Jet Tray*   Deaerator by Cochrane Environmental Systems.

**jetty**   A structure that extends into an open body of water to influence currents or tides or protect a harbor.

*Jet-Wet*™   Dry polymer feeding system by Fluid Dynamics, Inc., and Allied Colloids, Inc.

*Jigrit*   Screw-type grit washer by Jeffrey Division/Indresco.

*J-Mate*   Combination heating and shearing device for sludge volume reduction by JWI, Inc.

*Johnson Filter*   Former name of Wheelabrator Engineered Systems, Inc., Johnson Screens.

*Johnson Screen*   Wedgewire screen media by Wheelabrator Engineered Systems, Inc., Johnson Screens.

**journal bearing**   A cylindrical bearing that supports a rotating shaft.

*J-Press®*   Plate and frame filter press by JWI, Inc.

*J-Track™*   Nonmetallic return track for chain and flight sludge collectors by FMC Corp.

**JTU**   See "Jackson turbidity unit."

*JUD*   Belt tracking system for belt filter press by Klein America, Inc.

**junction box**   An enclosure within which electrical cables are connected and/or terminated.

# K

*Kaldnes System*   Biological wastewater treatment system by Purac Engineering, Inc.

*KAMET*   Municipal effluent treatment system by Krofta Engineering Corp.

*Kan-Floc*™   Wastewater flocculant/coagulant by Kem-Tron.

**kaolin**   A type of fine, white clay material.

**katabatic wind**   A localized wind that flows down valley or mountainous slopes, usually at night, caused by the descent of cold air as the valley slopes undergo rapid nocturnal cooling.

*Katec*®   Thermal oxidizer by Grace TEC Systems.

*Kat-Floc*™   Wastewater flocculant/coagulant by Kem-Tron.

*Katox*   Catalytic oxidizer by Adwest Technologies, Inc.

*KD-HF*™   Deionization system by Kinetico Engineered Systems, Inc.

*Kebab*™   Disc-type oil skimmer by Vikoma International.

*Keene*   Former equipment manufacturer acquired by Amwell, Inc.

*Kenics*®   Static mixer product line by Chemineer, Inc.

*Kenite*®   Diatomite material by Celite Corp.

*Key-Tech*   Water and wastewater purification system product line by Keystone Engineering & Treatment Technology Co.

*K-Floc*™   Wastewater flocculant/coagulant by Kem-Tron.

*K-Floor*   Suspended monolithic filter floor by PWT Americas.

**kg**   See "kilogram."

**kiln**   A heated enclosure for processing a substance by drying or burning.

**kiln dust**   See "cement kiln dust."

**kilogram (kg)**   A unit of mass equivalent to 1000 grams or approximately 2.205 lb.

**kilowatt (kW)**   A measure of power equal to 1000 watts. One horsepower equals 0.746 kW.

**kilowatt-hour (kWh)**   A unit of energy equal to that expended by 1 kW in 1 hr.

**kinematic viscosity**   A fluid's absolute viscosity divided by its mass density.

**kinetic energy**   Energy that is possessed by a body of matter as a result of its motion.

**KIWA**   Netherlands Waterworks Testing and Research Institute.

*Klampress®*   Belt filter press by Ashbrook Corp. (U.S.) and Simon-Hartley, Ltd. (U.K.).

*Kleer Flow*   Spiral wound reverse osmosis membranes by Great Lakes International, Inc.

*Klenphos-300*   Zinc phosphate corrosion inhibitor by Klenzoid, Inc.

*Klensorb*   Oil and grease absorbent by Calgon Carbon Corp.

*KL Series™*   Cleaning powders for removal of membrane foulants by King Lee Technologies.

*K²Modular™*   Volumetric screw feeder by K-Tron North America.

*Koagulator*   Solids contact clarifier by Zimpro Environmental, Inc.

*Koch Corrosion Control*   Former name of KCC Corrosion Control Co.

*Koflo®*   In-line static mixers by Koflo Corp.

*Komara*™   Floating oil skimmer by Vikoma International.

*Kompress*®   Belt filter press by Komline-Sanderson.

*Komprimat*   Fish/screenings separation system by Hellmut Geiger GmbH & Co.

*Konsolidator*   Wet scrubber solids filter by CMI-Schneible Co.

*Koppers*   Former manufacturer of traveling bridge filters; the product line was acquired by Infilco Degremont, Inc.

*Koro-Z*   Polyvinyl chloride biological filter media formerly offered by B.F. Goodrich Co.

**kraft**   An alkaline chemical pulping process, using salt cake as makeup.

*Kraus-Fall*   Peripheral feed clarifier by Graver Co.

**krill**   A small, shrimp-like marine crustacean that is a major food source for whales, seals, and squid.

*Kruger/Fuchs*   Autothermal thermophilic aerobic digestion system by Krüger, Inc.

*KUBE*³   Belt filter press by Klein America, Inc.

**kW**   See "kilowatt."

**kWh**   See "kilowatt-hour."

*K-W Products*   Former equipment manufacturer acquired by Smith & Loveless, Inc.

# L

**L-10 life** See "B-10 life."

**LAER** Lowest achievable emission rate.

*Lagco* Parshall flume by F.B. Leopold Co., Inc.

**lagoon** An excavated basin or natural depression that contains water, wastewater, or sludge.

**laid length** The total length of a pipe or pipeline after it has been placed in position.

**lake** An inland body of water.

*Lakos IPC* Self-cleaning pump intake screen by Claude Laval Corp.

**LAL test** See "limulus amebocyte lystate test."

**Lamella®** Gravity settler/thickener using inclined plates by Parkson Corp.

*LamGard* Automated oxygen control system by Lamson Corp.

**laminar flow** A flow situation in which fluid moves in parallel layers, usually with a Reynolds number less than 2000.

*Lam-Pak®* Package treatment plant by Graver Co.

*Lancom™* Flue gas monitoring system by Land Combustion, Inc.

*Lancy™* Wastewater treatment product line offered by U.S. Filter Corp.

**land application** The disposal of wastewater or sludge onto land under controlled conditions.

**land ban** The provisions of the Resource Conservation and Recovery Act prohibiting land disposal of specific toxic materials unless they meet applicable treatment standards.

**land disposal**  Application of wastewater sludges to the soil without production of usable agricultural products.

**landfarming**  Application of organic waste onto surface soil for the purpose of controlled biodegradation.

**landfill**  A land disposal site that employs an engineering method of solid waste disposal to minimize environmental hazards and protect the quality of surface and subsurface waters.

**Langelier Saturation Index (LSI)**  A measure of the degree of saturation of calcium carbonate in water based on pH, alkalinity, and hardness. A positive LSI indicates that calcium carbonate may precipitate from solution to form scale.

*Lanpac*  Packing media for air stripping towers by Lantec Products, Inc.

**lapse rate**  The rate at which temperature declines as altitude rises.

*L\*ARO*  Reverse osmosis system by the former L\*A Water Treatment Corp.

*Lasaire®*  Lagoon aeration system by A.B. Marketech, Inc.

**latent heat**  The heat required to cause a change of state at constant temperature, such as the vaporization of water or the melting of ice.

**lateral**  A secondary pipe that extends from a main water pipe or header.

*Lateral Flow Sludge Thickener™*  Gravity sludge thickener by Gravity Flow Systems, Inc.

**lateral sewer**  A sewer that connects the collection main to the interceptor sewer.

**launder**  A trough used to transport water.

**laundry wastes**  Wastewater from industrial laundries that may be characterized by the presence of lint, fibers, oils, and greases.

*L\*A Water Treatment*   Former name of PWT Americas.

**LC₅₀**   See "lethal concentration."

**LCR**   Lead Copper Rule.

**LD₅₀**   See "lethal dose."

**LDRs**   Land disposal restrictions. EPA-promulgated rules implementing the land ban.

**leachate**   Fluid that percolates through solid materials or wastes and contains suspended or dissolved materials or products of the solids.

**leach field**   The area of land into which a septic tank drains or wastewater is discharged.

**lead**   A trace element and cumulative poison that may be inhaled or ingested in food or water. Chemical formula is Pb.

*Leadtrak*   Test kits used to determine lead content of water by Hach Company.

**leaf filter**   A precoat filter with flat elements or leaves.

**leakage**   (1) The presence of an ionic species in ion exchanger effluent that usually indicates bed exhaustion. (2) The uncontrolled loss of water from a tank or aquifer.

*Leakwise*®   Oil-on-water monitoring systems by Agar Corporation.

*L'eau Claire*®   Upflow filter products by U.S. Filter Corp.

*Lectra/San*   Marine sanitation system by Eltech International Corp.

*Ledward & Beckett*   Former screening equipment manufacturer acquired by Brackett Green, Ltd.

**left bank**   The bank to the left of a river or stream when facing downstream.

**Legionella**   A genus of bacteria, some species of which have caused a type of pneumonia called Legionnaires Disease.

*Lemna*®   Biological wastewater treatment system utilizing aquatic duckweed by Lemna Corp.

*Lemnaceae*   Floating aquatic plants that provide a habitat for aquatic organisms capable of metabolizing wastewater organics.

**lentic water**   Standing or stagnant pond, swamp, or marsh water.

*Leo-Lite*   Fiberglass effluent and scum trough by F.B. Leopold Co.

*LeoVision*   PC-driven graphics display of treatment plant operating conditions by F.B. Leopold Co.

**lethal concentration**   The concentration of a substance that is fatal to a specified percentage of the population, usually expressed at the 50% level as $LC_{50}$.

**lethal dose**   The quantity of a substance that is fatal to a specified percentage of the population, usually expressed at the 50% level as $LD_{50}$.

*Level Bed Agitator*   Agitator used in composting systems by Wheelabrator Clean Water Systems, Inc., Bio Gro Division.

*Level Mate*™   Level measurement instrument by Ametek.

*Lewatit*   Ion exchange resins by Miles.

**lichen**   A sponge-like plant growing on wood, stone, or soil that is formed by an association of a fungus and alga, and often used as an air pollution indicator species.

*Lifeserver*™   Built-in-place wastewater treatment plants by Davis Water & Waste Industries, Inc.

*Lift Screen*   Reciprocating rake bar screen by Envirex, Inc.

**lift station**   A chamber that contains pumps, valves, and electrical equipment necessary to pump water or wastewater.

*Lightspeed*   Digital fiber optic flowmeter by Presto-Tek Corp.

**lignin**   An organic substance that forms the chief part of wood tissue.

**lignite**  A type of coal with a low energy content; also called brown coal.

*Limberflo*  Precast filter bottom system by Aerators, Inc.

**lime**  The term generally used to describe ground limestone (calcium carbonate), hydrated lime (calcium hydroxide), or burned lime (calcium oxide).

**lime-and-settle**  Common term for treatment technologies that utilize chemical precipitation and sedimentation processes.

**lime kiln**  A unit used to calcine lime.

**lime recalcining**  Recovery of lime from water or wastewater sludge, usually with a multiple-hearth furnace.

**lime slaker**  A device used to hydrate quicklime.

**lime-soda softening**  The addition of sufficient lime and soda ash to raw water to achieve a reduction of carbonate and noncarbonate hardness.

**lime softening**  The addition of sufficient lime to raw water to achieve a reduction of carbonate hardness.

**lime stabilization**  The addition of lime to untreated sludge to raise the pH to 12 for a minimum of 2 hr to chemically inactivate microorganisms.

**limestone**  A sedimentary rock composed primarily of calcium carbonate.

**limnology**  The study of freshwater lakes and their flora and fauna.

**Limulus amebocyte lystate test (LAL test)**  A test used to determine presence of endotoxins in treated water, commonly used on water to be used for pharmaceutical purposes.

**liner**  (1) A barrier of plastic, clay, or other impermeable material that prevents leachate from contacting surface or subsurface water. (2) A protective, corrosion-resistant layer attached or bonded to the inside of a tank.

*Link-Belt*® Environmental equipment product line by FMC Corp., MHS Division.

**lipid** Water-insoluble molecules important in the structure of cell walls and membranes.

**lipophilic** Having an affinity for oil.

*Liquaclone*® Hydrocyclonic solids separation unit for removal of granular solids from liquid discharges by Sanborn Environmental Systems.

*Liquapac*™ Solids removal unit for spent coolant/oil clarification applications by Sanborn Environmental Systems.

**liquefaction** The process of making or converting a solid or gas to a liquid.

**liquid** The state of matter between the solid and gaseous states in which matter possesses a definite volume and flows freely, but has no definite shape.

*Liquid A*™ Sludge stabilization process by RDP Co.

**liquid chlorine** Chlorine compound that contains no water and results from gaseous chlorine under high pressure, which is stored in steel drums and cylinders.

**liquid-liquid extraction** See "solvent extraction."

*Liquidow*® Calcium chloride products by Dow Chemical Co.

*LiquidPure* Small, low-cost activated carbon drum adsorber by American Norit Company, Inc.

*Liqui-Fuge*™ Internally fed rotary fine screen by Vulcan Industries, Inc.

*Liqui/Jector*® Liquid/gas coalescer by Osmonics, Inc.

*LiQuilaz*® In-line sensor for particle measurement by Particle Measuring Systems, Inc.

*Liquiphant* Liquid level indicator by Endress+Hauser.

*Liqui-pHase*® Carbon dioxide neutralization system by Liquid Carbonic.

*Liquipure* Company acquired by U.S. Filter Corp.

***Liqui-Strainer***   Externally fed rotating drum screen by Vulcan Industries, Inc.

***LiquiTrak™***   Nonvolatile residue monitor by TSI Inc.

***Liquitron®***   Pump controller by Liquid Metronics, Inc.

**liquor**   An aqueous solution of one or more chemical compounds.

**listed hazardous waste**   The designation for a waste material that appears on an EPA list of specific hazardous wastes or hazardous waste categories.

**liter (L)**   A unit of volume equal to 1000 cubic centimeters, or 1.057 quarts.

**lithology**   The character or description of rocks in terms of their physical and chemical characteristics.

**lithosphere**   The solid portion of earth, composed of rocks and soil.

***Litmustik®***   Pocket pH tester by Omega Engineering, Inc.

**litter**   Solid waste or garbage from human activity deposited indiscriminately.

***Little Fox***   Modular wastewater treatment plant for marine applications by Red Fox Environmental, Inc.

**littoral zone**   The area of the shore line between high and low tides where rooted water plants can grow.

**LLE**   Liquid-liquid extraction. See "solvent extraction."

***LME®***   Inclined-plate separator by Zimpro Environmental, Inc.

***LMI***   Liquid Metronics, Inc.

***L&N***   Leeds & Northrup, a unit of General Signal.

**LNG**   Liquefied natural gas.

***Load Limitor***   Automatic chain tensioning system for traveling water screens by Envirex, Inc.

**LOAEL**   See "lowest-observed-adverse-effect level."

**loam soil**   A rich soil consisting of organic material, sand, silt, and clay.

*Lobe-Aire®*  Rotary lobe blower by Spencer Turbine Company.

*Lobeflo™*  Rotary lobe pump by MGI Pumps, Inc.

*Lobeline™*  Positive displacement pump by Positive Flow Systems, Inc.

**localized corrosion**  Corrosion taking place at a relatively high speed in limited sections of the area exposed to the corrosive medium.

*Lo-Cat®*  Hydrogen sulfide oxidation process for anaerobic bio-gas systems by Wheelabrator Clean Air Systems, Inc.

**LOEL**  Lowest-observed-effect concentration.

**log boom**  A floating structure of logs or timber used to protect an intake, dam, or other structure by deflecting floating material.

**log-death phase**  Bacterial growth phase where the microbe death rate exceeds the production rate of new cells.

*LogEasy™*  Particle counter by Hach Co.

**log-growth phase**  Bacterial growth phase where cells divide at a rate determined by their generation time and their ability to process food.

*Lo-Head™*  Traveling bridge filter by Agency Environmental, Inc.

**LOI**  Loss on ignition.

*Longopac*  Screenings bagging system by Spirac Engineering.

*LoNo$_x$™*  Combustion burner by John Zink Co.

*LOOP*  Package wastewater treatment process using an oxidation ditch process by Smith & Loveless, Inc.

*Loop Chain*  Nonmetallic, filament wound sludge collector chain by Envirex, Inc.

*Lo-Pro™*  Air stripper by ORS Environmental Systems.

**loss of head**   A decrease in head energy that results from a bend, obstruction, or expansion in a channel or pipeline.

**lotic water**   Rapidly flowing water of a river or stream.

**lowest-observed-adverse-effect level (LOAEL)**   The lowest dose of a substance to cause an increase in the frequency or severity of an adverse effect in an exposed population.

**LPG**   Liquefied petroleum gas.

*LSC*™   Package spray-type deaerating heater by Graver Co.

**LSI**   See "Langelier Saturation Index."

**LSTK**   Lump sum turnkey.

**LTA**   Low-temperature additive.

**LUST**   Leaking underground storage tank.

**LWL**   Low water level.

*Lyco*™   Wastewater treatment equipment product line by U.S. Filter Corp.

**lyse**   To undergo lysis.

**lysimeter**   A device used to measure or obtain samples of water draining through soil.

**lysis**   The rupture of a cell that results in loss of its contents.

# M

m³   See "cubic meter."

**macerate**   To chop or tear.

*MacPac*   Chemical feed package by Milton Roy Co.

*Macro-Cat*   Ion exchange resin by Sybron Chemicals, Inc.

**macroencapsulation**   Isolation of waste by embedding or surrounding it with a material that acts as a barrier between the waste and air, water, or other materials.

**macrofloc**   Destabilized floc particle that is too large to penetrate a granular media filter bed.

**macrofouling**   The biological fouling of a water system with macroorganisms including clams, barnacles, and mussels.

*Macrolite*™   Ceramic filter media by Kinetico Engineered Systems, Inc.

**macroorganisms**   All organisms larger than microscopic, visible to the unaided eye.

**macrophyte**   A type of macroscopic plant life.

**macroreticular**   A term describing a resin having a pore structure even after drying.

**macroscopic**   Capable of being seen with the naked eye.

**MACT**   See "maximum achievable control technology."

*Magicblock*™   Fluid control system by Osmonics, Inc.

*Magna*   Rotor aerator by Lakeside Equipment Corp.

*Magna Cleaner*™   Liquid cyclone by Andritz-Ruthner, Inc.

*MagneClear*™   Magnesium hydroxide water treatment products by Martin Marietta Magnesia Specialties, Inc.

*Magnifloc*®   Polyelectrolyte used to enhance liquid/solid separation by American Cyanamid Co.

*Magnum*   Ultraviolet disinfection equipment by Atlantic Ultraviolet Corp.

*Magnum®*   Belt filter press by Parkson Corp.

*Magnum™*   Catalytic oxidizer by Grace TEC Systems.

*Magox®*   Magnesium oxide by Premier Services Corp.

**make-up**   Fluid introduced in a recirculating stream to maintain an equilibrium of temperature, solids concentration, or other parameter(s).

**malathion**   A common organophosphate insecticide.

**M-alkalinity**   See "methyl orange alkalinity."

*Mallard*   Bridge-mounted clarifier scum removal system by Copa Group.

*Mammoth®*   Brush aerator by Zimpro Environmental, Inc.

**manganese greensand**   See "greensand."

*Manhattan Process*   High-rate filtration process by Roberts Filter Manufacturing Co.

**manhole**   An opening in a vessel or sewer to permit human entry. Also called manway.

**manifest**   A form used to identify the origin, quantity, and composition of a hazardous waste during its transportation from the point of generation to the point of final treatment or disposal.

**Manning's formula**   Formula used to measure flow in an open channel based on the cross-sectional area of the flowing stream and the hydraulic radius, slope, and roughness of the channel.

**manometer**   A u-tube device filled with a liquid used to measure pressure differentials in liquids or gases.

*Manor®*   Filter press by Simon-Hartley, Ltd.

*Manu-Matic*   Manual pressure leaf filter cleaning system by U.S. Filter Corp.

*Manver*   Chemical composition used in analysis of water hardness by Hach Co.

*Manville Filtration*   Former name of Celite Corp.

*MARD*   Motor-actuated rotary distributor for trickling filters by General Filter Co.

*Marox*   Pure oxygen wastewater treatment system by Zimpro Environmental, Inc.

*MARS*   Membrane-controlled biological wastewater treatment technology by Krüger, Inc.

**marsh gas**   Methane gas produced by the anaerobic decomposition of organic matter in wetland areas. Also called "swamp gas."

**marshland**   An area of soft, wet land vegetated by reeds and grasses.

**masking**   Blocking out or covering a sound or smell with another.

*Maspac®*   Plastic packing media by Clarkson Controls & Equipment Co.

**mass balance**   An analysis that delineates changes that take place in a reactor or system by quantifying system inputs and outputs.

**mass spectrometer**   An instrument used for the analysis of organic materials in environmental samples by sorting ions according to their masses and electrical charges.

*Master-Flo*   Bladder pump by American Sigma.

**material balance**   See "mass balance."

**Material Safety Data Sheet (MSDS)**   Data sheet containing descriptive information required by the Occupational Safety and Health Administration for hazardous materials.

**materials recovery facilities (MRFs)**   A central facility where recycled materials are prepared and sorted.

*Mather Platt*   Former name of the screening equipment division of Weir Pumps, Ltd.

**maturation pond**   An aerobic waste stabilization pond used for polishing treated wastewater effluent.

*MaxAir™*   Wide-band coarse-bubble diffuser by Environmental Dynamics, Inc.

*Maxi-Flo®*   Pressurized sand filter product line by U.S. Filter Corp.

*Maxim®*   Seawater conversion evaporator by Beaird Industries, Inc.

*MaxiMizer®*   Solid bowl centrifuge by Alfa Laval Separation, Inc.

**maximum achievable control technology (MACT)**   The level of air pollution control technology required by the Clean Air Act.

**maximum contaminant level (MCL)**   The maximum permissible level of a contaminant in water delivered to the free flowing outlet of the ultimate user of a public water system.

**maximum contaminant level goal (MCLG)**   The maximum level of a contaminant, including an adequate safety margin, at which no known or anticipated adverse effect on human health would occur.

*Maxipress*   Belt filter press by Envirex, Inc.

*Maxi-Rotor*   Rotary brush aerator by Krüger, Inc.

*Maxi-Strip®*   Hydraulic venturi for removal of volatile organic compounds by Hazleton Environmental Products, Inc.

*Maxi-Yield™*   Polymer blending system by Wallace & Tiernan, Inc.

*Max-Pak™*   Plastic packing media by Jaeger Products, Inc.

*Maz-O-Rator*   Solids grinder by Robbins & Myers, Inc.

**MBAS**   See "methylene blue active substance."

**MCC**   Motor control center.

*McGinnes-Royce*   Manufacturer of screening equipment whose product line was acquired by Envirex, Inc.

**MCL**   See "maximum contaminant level."

**MCLG**   See "maximum contaminant level goal."

*m-ColiBlue24*™   Laboratory coliform test broth by Hach Co.

**MCRT**   See "mean cell residence time."

**MDTOC**   Minimum detectable threshold odor concentration.

**me**   See "milliequivalent."

**ME**   Multiple effect. See "multiple effect distillation."

**mean cell residence time (MCRT)**   The average time that a microbial cell remains in an activated sludge system. It is equal to the mass of cells divided by the rate of cell wasting from the system.

**mean sea level**   The average sea level for all stages of the tide.

**mean velocity**   The average velocity of a fluid flowing in a channel, pipe, or duct, determined by dividing the discharge by the cross-sectional area of the flow.

**MEB**   Multiple effect boiling. See "multiple effect distillation."

**mechanical aeration**   The mechanical agitation of water to promote mixing with atmospheric air.

**mechanical coupling**   A pipe coupling that does not require threading.

**mechanical draft cooling tower**   A cooling tower that depends on fans for introduction and circulation of its air supply.

*MECO*®   Mechanical Equipment Co., Inc.

*Mectan*™   Grit chamber by John Meunier, Inc.

*MecTool*®   Remediation mixing system by Millgard Environmental Corp.

**MED**   See "multiple effect distillation."

**medical waste**   Any solid waste generated in the diagnosis, treatment, or immunization of humans or animals.

*Medina*®   Bioremediation products by Medina Products.

*Megatron*™   Ultraviolet water purification system by Atlantic Ultraviolet Corp.

*Megos*®   Ozone generation system by Capital Controls Co., Inc.

*Mellafier*   Inclined-plate clarifier by Industrial Filter & Pump Mfg. Co.

**melting point**   The temperature at which a solid changes to a liquid.

**melt water**   Water derived from the melting of ice and snow.

*Membio*®   Aerobic biological digester by Memtec America Corp.

*Membralox*®   Ceramic membrane filters by U.S. Filter Corp.

**membrane**   A thin barrier that permits passage of particles of a certain size or particular nature.

**membrane diffuser**   Fine-bubble aeration diffuser with perforated flexible plastic membranes.

**membrane filter**   A paper-like filter, with small pore sizes, that is capable of retaining bacteria for use in the laboratory examination of water.

**membrane processes**   Processes including reverse osmosis, electrodialysis, and ultrafiltration that use membranes to remove dissolved material or fine solids.

**membrane softening**   A water softening process that utilizes semipermeable membranes to remove hardness constituents such as calcium and magnesium from water.

*Membrastill*™   Pharmaceutical water system by U.S. Filter Corp.

*Memcor*® Continuous microfiltration system by Memtec America Corp.

*Memory-Flex*™ Check valve by Val-Matic, Corp.

*Mem\*Recon*™ Reverse osmosis membrane reconditioning product by King Lee Technologies.

*Memstor* Membrane storage agent by King Lee Technologies.

*Memtek* Memtek Product Group of Wheelabrator Engineered Systems, Inc.

*Memtrex*™ Pleated filters by Osmonics, Inc.

*Mensch*™ Reciprocating rake bar screen by Vulcan Industries, Inc.

**MEP** See "multiple extraction procedure."

**meq/L** See "milliequivalents per liter."

**mercaptans** Organic compounds, or thioalcohols (thiols), containing sulfur and noted for their disagreeable odor.

*Merco*® Centrifuge by Dorr-Oliver, Inc.

**mercury** A heavy metal element that when absorbed or ingested by humans is excreted from the body very slowly and can be lethal in very low concentrations.

*Merlin*® Progressing cavity pump by MGI Pumps, Inc.

*Mer-Made* Filter leaves for vacuum diatomite filters by Mer-Made Filter, Inc.

*Mesa-Line*® Portable submersible pump by Crane Pumps & Systems.

**mesh** The number of openings per lineal inch, measured from the center of one wire or bar to a point 1 in. (25.4 mm) distant.

**mesophiles** Bacteria that grow best at temperatures between 25 and 40°C.

**mesophilic digestion** Anaerobic sludge digestion within a mesophilic range of approximately 25–40°C.

**mesophyte** A plant that grows under typical or moderate amounts of atmospheric water supply.

**mesosphere**   The level of the earth's atmosphere that exists above the stratosphere.

**mesotrophic lake**   A lake between the oligotrophic and eutrophic stages that remains aerobic, although a substantial depletion of oxygen has occurred in the hypolimnion.

**metabolize**   The biological conversion of organic matter to cellular matter and gaseous byproducts.

**metal**   In general, those elements that easily lose electrons to form positive ions.

*Metal-Drop*™   Flocculant/coagulant by Kem-Tron.

**metal finishing wastes**   Wastewater from electroplating, galvanizing, and other metal finishing operations that may be characterized by the presence of acids, caustics, and metal contaminants.

**metal salt coagulants**   Salts of alum and iron (III) commonly used as water treatment coagulants.

*MetalWeave*®   Stainless steel fabric tank baffles by Eimco Process Equipment Co.

**metering pump**   A pump used to provide controlled injection of a chemical additive into a fluid flow.

**methane**   A colorless, odorless combustible gas that is the principal byproduct of anaerobic decomposition of organic matter in wastewater. Chemical formula is $CH_4$.

**methane formers**   See "methanogens."

**methanogens**   Group of anaerobic bacteria responsible for conversion of organic acids to methane gas and carbon dioxide.

**methanol**   A solvent often used as a supplemental carbon source during denitrification. Chemical formula is $CH_3OH$.

**methemoglobinemia**   Disease occurring primarily in infants who have ingested water high in nitrates. Also called "blue baby syndrome."

**methylene blue active substance (MBAS)**   Anionic surfactants that react with methylene blue to form a chloroform soluble complex.

**methyl orange**   A color indicator used in acid and base titrations.

**methyl orange alkalinity**   A measure of the total alkalinity of an aqueous solution determined through titration with a methyl orange color indicator.

**metric ton**   A unit of mass equal to 1000 kg or approximately 2204 lb. Also called "tonne."

**mg**   See "milligram."

**mgd**   Million gallons per day.

**mgid**   Million gallons (imperial) per day.

**mg/L**   See "milligrams per liter."

**mg/L as CaCO₃**   See "calcium carbonate equivalent."

**mho**   Unit of measurement for conductivity equal to the reciprocal of resistivity (ohm).

*MHT*®   Magnesium hydroxide by Dow Chemical Co.

*Micrasieve*™   Pressure-fed fine static screen by Andritz-Ruthner, Inc.

*Micro/2000*®   Residual chlorine, chlorine dioxide, and potassium permanganate analyzer by Wallace & Tiernan, Inc.

**microbe**   Organism observable only through a microscope.

**microbiocide**   See "biocide."

*Microbloc*   Carbon bed volatile organic compound control system by Wheelabrator Clean Air Systems, Inc.

*Microcat*®   Microbial additive for use in biological wastewater treatment by Bioscience, Inc.

*Microchem*™   Tablet chlorinator by Mooers Products, Inc.

*MicroDAF*   Dissolved-air flotation by the former Princeton Clearwater.

**microelectronic water**   See "electronic-grade water."

**microencapsulation**   Isolation of a waste material by mixing it with a material that then cures or converts to a solid, nonleaching barrier.

**microfauna**   Animals not visible to the naked eye.

*MicroFID*™   Volatile organic compound detection unit by Photovac, Inc.

**microfiltration**   A low-pressure membrane filtration process that removes suspended solids and colloids larger than 0.1 microns.

*Micro Fine*   Ultrafiltration system by Memtec Americas Corp.

*Microfloat*™   Surface-mounted aspirating aerator for dispersed-air flotation by Aeration Industries, Inc.

**microfloc**   Destabilized floc particle that permits in-depth penetration of a granular media filter bed to optimize the filter's solid retention capacity.

*Microfloc*®   Product group of Wheelabrator Engineered Systems, Inc.

**microfouling**   The biological fouling of a water system with microorganisms including algae, fungi, and bacteria.

**microgram (µg)**   A unit of mass equal to one-millionth of a gram.

**micrograms per cubic meter (µg/m³)**   A measure of the concentration of particulate or gaseous matter in air, commonly used in reporting air pollution data.

**microirrigation**   A water management irrigation technique using a microsprinkler or drip irrigation system to minimize water runoff.

*Micro-Matic*®   Rotating microscreen by U.S. Filter Corp.

*Micromesh Strainer*   Microscreen by Lakeside Equipment Co.

**micrometer (μm)**   See "micron."

**micromho**   A unit measure of conductivity equal to one-millionth of a mho.

**micron (μ)**   A unit of length equal to one-millionth of a meter. Also called "micrometer."

*Micronizer™*   Fine-bubble dissolved-air flotation device by the former Microlift Systems, Inc.

**microorganism**   Organisms observable only through a microscope. Also called "microbes."

*Micro-Pi™*   Pressure-fed rotary screen by Andritz-Ruthner, Inc. (Western Hemisphere) and Contra-Shear Engineering, Ltd.

*Micro-Polatrol®*   Cathodic protection power units by Cathodic Protection Services Co.

**microscope**   An instrument used for visual magnification of small objects.

**microscreen**   A surface filtration device consisting of a rotating drum with a fine-mesh screen fixed to its periphery. As water flows through the interior of the drum, solids are retained by the mesh for removal by a high-pressure spray wash.

*Micro-Sieve*   Microscreen formerly offered by Passavant Corp.

**microstrainer**   See "microscreen."

*Micro-T*   Turbidimeter with remote station monitoring by HF Scientific, Inc.

*Microtox®*   Acute toxicity test by Microbics Corp.

**midge fly**   An insect that may infest a water system and whose larvae, known as "blood worms," feed on algae, protozoans, and decaying vegetation.

*Midi-Rotor*   Rotary brush aerator by Krüger, Inc.

**midnight dumping**   The deliberate and illegal disposal of sludge or other waste materials at an unauthorized, nonpermitted location.

*MightyPure*™   Ultraviolet water purification system by Atlantic Ultraviolet Corp.

**MIL**   Military specification.

*MIL*   Prop and specialty agitator product line by Denver Equipment Co.

**mill**   To grind or crush.

**milliequivalent (me)**   One-thousandth of an equivalent weight.

**milliequivalents per liter (meq/L)**   An expression indicating the concentration of a solute, which is calculated by dividing the concentration, in mg/L, by the equivalent weight of the solute.

**milligram (mg)**   A unit of mass equal to one-thousandth of a gram.

**milligrams per liter (mg/L)**   A common unit of measurement of the concentration of a dissolved material in solution.

**milliliter (mL)**   A unit of volume equal to 1 $cm^3$.

*Milliscreen*™   Internally fed rotary fine screen by Andritz-Ruthner, Inc. (Western Hemisphere) and Contra-Shear Engineering, Ltd.

**mill scale**   An oxide coating formed on steel when heated in connection with hot working or heat treatment.

**mineral**   Any naturally occurring material having a definite chemical composition and structure.

**mineral acidity**   Acidity caused by the presence of mineral acids.

**mineral acids**   Inorganic acids including hydrochloric, nitric, and sulfuric acid.

**mineralization**   The conversion of an organic material to an inorganic form by microbial decomposition.

**mineral water** Bottled water with at least 250 parts per billion dissolved solids, from a source tapped at one or more bore holes or springs, originating from a geologically and physically protected underground water source.

*MiniBUS*™ Ultrafiltration membrane cartridge by Cuno Separations Systems Division.

*MiniChamp* Chemical induction unit by Gardiner Equipment Co., Inc.

*Minigas*® Multigas detector by Neotronics.

*Mini-Magna* Rotor aerator by Lakeside Equipment Corp.

*Minimax* Dewatering pressure filter by Larox, Inc.

*Mini-Maxi* Dissolved-air flotation unit by Tenco Hydro, Inc.

*Mini-Miser*™ Multiple-feed dewatering system by Recra Environmental.

*Mini Monster*® Low-flow sewage grinder by JWC Environmental.

*Mini Osec* Electrolytic chlorination system by Wallace & Tiernan.

*Minipure*™ Ultraviolet water purification system by Atlantic Ultraviolet Corp.

*Mini-Ring* Plastic biological filter media by Mass Transfer, Inc.

*Mini-San* Tablet feeder disinfection system by Eltech International Corp.

*Miniseries*™ Packaged desalination plant by Matrix Desalination, Inc.

**miscible** Capable of being mixed.

**mist eliminator** A device used to remove entrained droplets of water from a vapor stream produced during evaporation.

*Mist-Master*® Mesh pad mist eliminators by ACS Industries, Inc.

*Mixaerator* Static mixing aerators by Ralph B. Carter Co.

*Mixco*   Batch mixer by Lightin.

**mixed bed demineralizer**   Ion exchange demineralizer containing strong-acid and strong-base resins in a single vessel.

**mixed liquor**   The mixture of wastewater and activated sludge undergoing aeration in the aeration basin.

**mixed liquor suspended solids (MLSS)**   Suspended solids in the mixture of wastewater and activated sludge undergoing aeration in the aeration basin.

**mixed liquor volatile suspended solids (MLVSS)**   The volatile fraction of the mixed liquor suspended solids.

**mixed-media filter**   Granular media filter utilizing two or more types of filter media of different sizes and specific gravities, usually silica sand, anthracite, and ilmenite or garnet.

**mL**   See "milliliter."

**MLSS**   See "mixed liquor suspended solids."

**MLVSS**   See "mixed liquor volatile suspended solids."

*MobileFlow*   Trailer-mounted water treatment system by Ecolochem, Inc.

*MobileRO*   Trailer-mounted reverse osmosis system by Ecolochem, Inc.

*Mobius*   Fine-mesh belt screen by Pro-Ent, Inc.

**modeling**   A quantitative or mathematical simulation that attempts to predict or describe the behavior or relationships that result from a physical event.

*Mod-U-Flo*   Round-bottom clarifier by Western Filter Co.

*Moduflow*   Concrete gravity sand filter by Smith & Loveless, Inc.

*Modulab*®   High-purity water systems by U.S. Filter Corp.

*Modular Aquarius*®   Modular water treatment plant by Wheelabrator Engineered Systems, Inc., Microfloc Products.

*Modu-Plex*    Wet well pump station by Smith & Loveless, Inc.

**molality**    The number of moles of solute per kilogram of solvent.

**molarity**    The number of moles of solute per liter of solution.

**mole**    (1) The molecular weight of a substance containing an Avogadro's number of atoms or molecules. (2) A massive harborwork, breakwater, or jetty.

**molecular weight**    The weight of a molecule that may be calculated as the sum of the atomic weights of its constituent atoms.

*Molyver*    Reagent chemicals used to determine molybdenum concentration in water by Hach Company.

**monel**    A nickel alloy containing approximately 30% copper and having good mechanical and corrosion-resistant properties.

**monitoring well**    A well used to obtain samples for analysis or to measure groundwater levels.

*Monkey Screen*    Reciprocating rake bar screen by Brackett Green.

*Mono®*    Pump products by Ingersoll-Dresser Pump (U.S.) and H2O Waste-Tec (U.K.).

*Monoblock*    Carbon bed volatile organic compound control system by Wheelabrator Clean Air Systems, Inc.

*Monocluster*    Package water treatment plant by Graver Co.

**Monod equation**    A mathematical equation that describes the relationship between biomass production and the concentration of growth-limiting substrate.

*Mono-Ferm*    Iron and manganese removal gravity filter by Graver Co.

**monofill**    A solid waste disposal facility containing only one type or class of waste.

*Monoflo®*   Screenings grinder by Ingersoll-Dresser Pump (U.S.), Dresser Pump, and H2O Waste-Tec (U.K.).

*Monoflo*   Progressing cavity pump by MGI Pumps, Inc.

*Mono-Floc®*   Gravity sand filter with coagulant feed system by Graver Co.

*Monoflor®*   Cast-in-place filter underdrain by Infilco Degremont, Inc.

*Monolift®*   Vertical, progressing cavity groundwater pump by Ingersoll-Dresser Pump (U.S.) and H2O Waste-Tec (U.K.).

**monomedia filter**   A granular media filter utilizing a single size and type of filter media.

**monomer**   The basic molecule of a synthetic resin or plastic.

*Mono-Pak*   Concrete gravity filter by Graver Co.

*Mono-Pilot®*   Coagulant control center using a pilot filter column by Wheelabrator Engineered Systems, Inc., Microfloc Products.

*Monorake*   Traveling bridge raking mechanism for rectangular clarifiers by Dorr-Oliver, Inc.

*Mono-Scour®*   High-solids gravity sand filter by Graver Co.

*MonoSparj*   Coarse-bubble diffuser by Walker Process Equipment Co.

*Monosphere™*   Ion exchange resin by Dow Chemical Co.

*Monovalve® Filter*   Gravity sand filter by Graver Co.

*Monozone®*   Ozone generation system by Capital Controls Co., Inc.

**Montreal Protocol**   A 1987 international agreement to phase out chlorofluorocarbons and replace them with hydrofluorocarbons. The full agreement name is the "Montreal Protocol on Substances That Deplete the Ozone Layer."

**most probable number (MPN)** Statistical analysis technique based on the number of positive and negative results when testing multiple portions of equal volume.

**mother liquor** The concentrated solution that remains after evaporation or crystallization.

**motive steam** High-pressure steam used to operate a steam-jet ejector or thermocompressor.

*MotoDip* Motorized slotted skimmer pipe by Walker Process Equipment Co.

*Mountain Filter Media* Former name of Sierra Silica, Inc.

**moving bed filter** A granular media filter that continuously cleans and recycles filter media while the filter continues to operate.

*Moyno®* Progressing cavity pump line by Robbins & Myers, Inc.

**MPA** Microscopic particulate analysis.

**MPN** See "most probable number."

*MPVT™* Multipurpose vertical turbine pump by Patterson Pump Co.

*MRF Press* Belt filter press by Idreco USA, Ltd.

**MRFs** See "materials recovery facilities."

*M-roy* Metering pump products by Milton Roy Co.

*MS Diffuser* Medium-bubble diffuser by Enviroquip, Inc.

**MSDS** See "Material Safety Data Sheet."

**MSF** See "multistage flash evaporation."

**MSF-BR** Multistage flash evaporation, brine recirculation.

**MSF-OT** Multistage flash evaporation, once-through.

**MSW** Municipal solid waste.

**MSWLF** Municipal solid waste landfill.

*MT®* Reverse osmosis membrane cleaners by B.F. Goodrich Co.

**MTBE** Methyl-tertiary-butyl-ether.

**MTBF** Mean time between failures.

*MTS* Mass Transfer Systems, Inc.

*MTS®* Trailer-mounted mobile treatment system for industrial wastewater treatment applications by Graver Co.

**MTTR** Mean time to repair.

**MUD** Municipal Utility District.

**mud balls** Agglomerations of floc, solids, and filter media in a filter bed that may grow into a larger mass and reduce filtration efficiency.

**mud flat** A muddy, flat low-lying tidal area.

**mud valve** A valve used to drain sediment from the bottom of a sedimentation basin.

*Muffin Monster®* Sewage grinder by JWC Environmental.

**mulch** A protective ground covering of compost, wood chips, sawdust, or other organic matter.

*Multdigestion* Two-stage digestion system formerly offered by Dorr-Oliver, Inc.

**multiclone** A set of individual cyclone separators arranged in parallel to remove particulate matter from air emissions.

*Multicoil* Indirect sludge drying system by Kvaerner Eureka, USA.

*Multicone* Aluminum induction cascade aerator used to strip gases or aerate water supply by Infilco Degremont, Inc.

*Multicrete®* Monolithic filter underdrain system by General Filter Co.

*MultiDraw* Circular clarifier with pumped suction sludge removal system using multiple nozzles by Walker Process Equipment Co.

*Multiflo* Flow distribution nozzle for rotary distributors by Amwell, Inc.

*Multiflo®*   Pump column used with jet aeration system by Jet Tech, Inc.

*MultiFlow*   Skid-mounted water treatment system by Ecolochem, Inc.

*Multi-Flow*   Polyvinyl chloride biological filter media formerly offered by B.F. Goodrich Co.

*Multilogger*   Water instrumentation device for measuring multiple parameters by Stevens Water Monitoring Systems.

*Multi-Mag™*   Electromagnetic flowmeter by Marsh-McBirney, Inc.

**multimedia filter**   Granular media filter utilizing two or more types of filter media of different sizes and specific gravities, usually silica sand, anthracite, and ilmenite or garnet.

**multiple effect distillation**   A thin-film evaporation process where the vapor formed in a chamber, or effect, condenses in the next, providing a heat source for further evaporation.

**multiple extraction procedure (MEP)**   Procedure used to simulate the leaching that a waste will undergo form repetitive precipitation of acid rain on a material.

**multiple-hearth furnace**   A furnace or incinerator consisting of numerous hearths that is used to incinerate organic sludges or recalcinate lime.

**multiple-stage flash evaporation**   See "multistage flash evaporation."

*Multi-Point*   Level controller by Drexelbrook Engineering Company.

*Multiport Valve™*   Valve used to manipulate filter backwashing and rinsing by U.S. Filter Corp.

*MultiRanger*   Level and volume measurement device by Milltronics, Inc.

**multistage flash evaporation (MSF)**   A distillation process in which a stream of brine flows through the bottom of chambers, or stages, each operating at a successively lower pressure, and a proportion of it flashes into steam and is then condensed.

*Multi-Tech®*   Chemical feed, contact flocculation, and filtration process by Culligan International Corp.

*Multi-Turi®*   Wet scrubber with high-energy venturi by CMI-Schneible Co.

*Multiwash®*   Sand filtration process using combined air/water backwash by General Filter Co.

*Multi-Wash®*   Wet scrubber by CMI-Schneible Co.

*Multi-Zone*   Anaerobic digestion system by Zimpro Environmental, Inc.

*Muncher®*   Sewage grinder products by Ingersoll-Dresser Pump (U.S.) and H2O Waste-Tec (U.K.).

*Munchpump®*   Packaged pump/grinder assembly by Ingersoll-Dresser Pump (U.S.) and H2O Waste-Tec (U.K.).

**municipal waste**   The combined solid and liquid waste from residential, commercial, and industrial sources.

**municipal wastewater treatment plant**   Treatment works designed to treat municipal wastewater.

*Muniflo*   Positive displacement rotary lobe sludge pump by Envirotech Company.

*Munox*   Selective bacteria for wastewater treatment by Osprey Biotechnics.

*Münster*   Trash rake cleaning mechanism by Landustrie Sneek B.V.

**muntz metal**   A brass containing approximately 60% copper and 40% zinc.

**muriatic acid**   Chemical formula is HCl; also known as "hydrochloric acid."

**MUS**   Minimum ultimate strength.

*Mushroom Ventilator*   Cast iron air diffuser by the former Knowles Mushroom Ventilator Co.

**mutagen**   A material that causes genetic change when it interacts with a living organism.

**mutagenic**   A chemical or agent with properties that cause mutation or disfiguring.

*Mutrator*®   Packaged pumping/grinding station by Ingersoll-Dresser Pump (U.S.) and H2O Waste-Tec (U.K.).

**MVC**   Mechanical vapor compression. See "vapor compression evaporation."

**MVR**   Mechanical vapor recompression. See "vapor compression evaporation."

**MW**   Megawatt.

*Mystaire*®   Air scrubber systems by Misonix, Inc.

# N

**NAAQS**  See "national ambient air quality standards."

**NACE**  National Association of Corrosion Engineers.

*NAFCO*  Fibrous precoat filter aid by Liquid-Solids Separation Corp.

**nanofiltration**  A specialty membrane filtration process that rejects solutes larger than approximately 1 nanometer (10 angstroms) in size.

**NAPAP**  National Acid Precipitation Assessment Program.

**NAPC**  National Air Pollution Control Association.

**NAPL**  Nonaqueous phase liquid.

*Nara*  Paddle dryer/processor by Komline-Sanderson.

*Nasty Gas*™  Regenerative blowers to move noxious or other exotic gases by EG&G Rotron, Inc.

**national ambient air quality standards (NAAQS)** U.S. standards established under the Clean Air Act to set limits on criteria pollutant levels in ambient (outdoor) air.

**National Contingency Plan (NCP)**  U.S. Federal regulations promulgated to implement the Comprehensive Environmental Response, Compensation and Liability Act and the Clean Water Act.

**National Emission Standards for Hazardous Air Pollutants (NESHAP)**  U.S. standards established under the Clean Air Act to set limits on pollutants that may pose an immediate hazard to human health.

**National Environmental Policy Act (NEPA)**  A 1969 U.S. Public Law declaring a national policy that encourages productive and enjoyable harmony between people and their environment to enrich their understanding of ecological systems and natural resources.

*National Hydro*   Former equipment manufacturer acquired by Amwell, Inc.

**National Pollutant Discharge Elimination System (NPDES)**   U.S. program to issue, monitor, and enforce pretreatment requirements and discharge permits under the Clean Water Act.

**National Priorities List (NPL)**   U.S. Federal list of hazardous waste sites addressed by the Comprehensive Environmental Response, Compensation and Liability Act.

**natural draft cooling tower**   A cooling tower in which the air flow through the tower occurs naturally, rather than mechanically, as a result of tower design.

**natural gas**   A naturally occurring mixture of hydrocarbon gases found in geologic formations beneath the earth's surface whose principal constituent is methane.

**natural organic matter (NOM)**   Term used to described the organic matter present in natural waters.

**natural resource**   An area, material, or organism useful to man.

*Nautilus®*   Traveling bridge siphon sludge collection system by Wheelabrator Engineered Systems, Inc., Microfloc Products.

**NBOD**   See "nitrogenous oxygen demand."

**NBS**   National Bureau of Standards.

**NCCLS**   National Committee for Clinical Laboratory Standards.

**NCP**   See "National Contingency Plan."

**NDP**   See "net driving pressure."

**NDT**   Nondestructive testing.

**neat solution**   Full-strength, undiluted solution.

**NEC**   National Electrical Code.

**needle valve**   A valve that controls flow by means of a tapered needle that extends through a circular outlet.

**negative head**   Filter operating condition that occurs when the pressure in the filter bed is below atmospheric pressure during a filter cycle.

**NELAP**   National Environmental Laboratory Accreditation Program.

**NEMA**   National Electrical Manufacturers Association.

**nematode**   An unsegmented, often parasitic worm.

*Nemo®*   Progressive cavity pump and macerator product line by Netzsch, Inc.

**neoprene**   A synthetic elastomer that is chemically, physically, and structurally similar to natural rubber.

*Neosepta®*   Electrodialysis membrane stack supplied by Graver Co.

*Neozone™*   Ozone generator by North East Environmental Products, Inc.

**NEPA**   See "National Environmental Policy Act."

**nephelometric turbidity unit (NTU)**   Unit of measure used in the measurement of turbidity by instrumentation.

*Neptune*   Former name of Microfloc Products group of Wheelabrator Engineered Systems, Inc.

**NESHAP**   See "National Emission Standards for Hazardous Air Pollutants."

**Nessler tubes**   Color comparison tubes used in making colorimetric measurements.

**net driving pressure (NDP)**   The net feed pressure of a reverse osmosis system plus the osmotic pressure of the permeate, minus the permeate line pressure and osmotic pressure of the feedwater.

**net head**   The head available for production of energy in a hydroelectric plant after deduction of all frictional losses.

**net positive suction head (NPSH)**   The difference between the total pressure head and the vapor pressure of the liquid being pumped.

*Net-Waste*   Screw press by Diemme USA.

*Neutralite*   Filter media used to neutralize acidic waters by U.S. Filter Corp.

**neutralization**   The chemical process that produces a solution that is neither acidic nor alkaline.

*Neutralizer Plus*™   Media for pH adjustment by Matt-Son, Inc.

*Neva-Clog*   Filter media of perforated metallic sheets by Liquid-Solids Separation Corp.

**New Source Performance Standards (NSPS)**   Standards established under the Clean Air Act to impose federal technology-based control requirements on emissions from new stationary sources of pollution.

**new water**   Water from any discrete source, such as a river, creek, lake, or well, that is deliberately brought into a plant site.

**NF**   See "nanofiltration."

**NFPA**   National Fire Protection Association.

*Nichols*   Former furnace manufacturer acquired by Hankin Environmental Systems, Inc.

**night soil**   Human fecal wastes spread on fields as fertilizer.

**NIH**   National Institute of Health.

**NIMBY**   "Not in my backyard." A common expression that indicates a preference for waste disposal or treatment to occur at some distant location.

**NIOSH**   National Institute for Occupational Safety and Health.

**NIST**   National Institute of Standards and Technology.

*Nitox®*   Activated carbon adsorbers by TIGG Corp.

*Nitra-Select™*   Selective nitrate removal media by Matt-Son, Inc.

**nitrate**   A stable, oxidized form of nitrogen having the formula $NO_3^-$.

**nitrate formers**   See "Nitrobacter."

*Nitraver*   Reagent chemicals used to determine nitrite concentration of solutions by Hach Company.

**nitric acid**   A strong mineral acid having the chemical formula $HNO_3$.

**nitrification**   Biological process in which ammonia is converted first to nitrite and then to nitrate.

**nitrite**   An unstable, easily oxidized nitrogen compound with the chemical formula $NO_2^-$.

**nitrite formers**   See "Nitrosomonas."

**Nitrobacter**   Nitrifying bacteria that convert nitrites to nitrates. Also called "nitrate formers."

**nitrogen cycle**   A graphical presentation of nitrogen's natural cycle from living animal matter through dead organic matter and back to living matter.

**nitrogen dioxide**   A reddish brown gas that usually results from a combustion process and is one of the primary air pollutants; it causes respiratory irritation and illness in relatively low concentrations. Chemical formula is $NO_2$.

**nitrogen fixation**   The conversion of atmospheric nitrogen into nitrogen compounds through biological activity.

**nitrogen, nitrate**   See "nitrate."

**nitrogen, nitrite**   See "nitrite."

**nitrogenous biochemical oxygen demand (NBOD)**
The portion of biochemical oxygen demand where oxygen consumption is due to the oxidation of nitrogenous material, measured after the carbonaceous oxygen

demand has been satisfied. Also called "second-stage BOD."

**nitrogenous BOD**   See "nitrogenous oxygen demand."

**nitrogenous oxygen demand (NOD)**   That portion of the oxygen demand associated with the oxidation of nitrogenous material, usually measured after the carbonaceous oxygen demand has been satisfied.

**nitrogen oxides (NOx)**   Compounds formed and released primarily by the burning of fossil fuels.

*Nitroseed*   Nitrifying toxicity screening test by Polybac Corp.

**Nitrosomonas**   Nitrifying bacteria that convert ammonia to nitrites under aerobic conditions and derive their energy from the oxidation. Also called "nitrite formers."

*Nitrox*™   ORP system by United Industries, Inc.

**NMO**   Nonmethane organic compound.

**NMR**   See "nuclear magnetic resonance."

**NNI**   See "noise and number index."

**NOAA**   National Oceanic and Atmospheric Administration.

**NOAEL**   See "no-observed-adverse-effect level."

**Nocardia**   Bacteria that can accumulate to create a nuisance foam in aeration basins and secondary clarifiers.

*No-Cling*   Traveling water screen media insert by Norair.

**NOD**   See "nitrogenous oxygen demand."

**no detectable emissions**   An atmospheric discharge with a concentration less than 500 parts per million by volume as measured by an appropriate detection instrument.

**nodulizing kiln**   See "calciner."

**NOEC**   No-observed-effect concentration.

**no-effect level**   See "no-observed-adverse-effect level."

**noise**   Any unwanted sound, independent of volume.

**noise and number index (NNI)**  An index for assessing air traffic noise based on the average perceived decibel level of air traffic and the number of aircraft heard.

**noise-induced hearing loss**  A hearing loss, or permanent threshold shift, resulting from noise exposure rather than the normal loss attributed to age.

*Nokia*  Fine-bubble diffusers by Munters.

**NOM**  See "natural organic matter."

**nonattainment area**  A geographic area in which the level of a criteria air pollutant is higher than the level allowed by federal standards.

**noncarbonate hardness**  The hardness in water caused by chlorides, sulfates, and nitrates of calcium and magnesium.

**noncombustible refuse**  Solid wastes that will not burn in a conventional incinerator.

**noncommunity water system**  A public water system that serves a nonresident population such as a campground, school, factory, or residence.

**noncondensable gas**  Gaseous material not liquefied when associated water vapor is condensed in the same environment.

**noncontact cooling water system**  A once-through cooling water system that does not come into contact with hydrocarbons or other wastewater, and is not recirculated through a cooling tower.

**nonionic polymer**  A polyelectrolyte with no net electrical charge.

**nonmetal**  Elements that hold electrons firmly and tend to gain electrons to form negative ions.

**nonpoint source**  A source, other than a point source, that discharges pollutants into the air or water.

**nonputrescible** Material that cannot be decomposed by biological methods.

**nonrenewable resource** A naturally occurring finite resource that cannot be renewed once it has been used.

**nonsettleable solids** Suspended solids that remain in suspension, usually for more than 1 hr.

**no-observed-adverse-effect level (NOAEL)** The maximum dose of a substance that produces no observed adverse effects.

*NOPOL*™ Bottom-mounted diffuser system by Aeration Industries, Inc.

*Noramer*® Water treatment polymers by Rohm & Haas.

*Norit Roz* Steam-activated, peat-based carbon product by Norit Americas, Inc.

**NORM** Naturally occurring radioactive materials.

**normality** A solution's relation to the "normal solution."

**normal solution** A solution that contains one equivalent weight of a substance per liter of solution.

*Nor-Pac*® Random tower packing by NSW Corp.

*Nortex* Side and boot seals for traveling water screens by Norair Engineering Corp.

*North*™ Internally fed rotating drum screen products by KRC (Hewitt) Inc.

*North American* Pressure leaf filter product line by Liquid-Solids Separation Corp.

*North Filter* Rotary fine screen by KRC (Hewitt) Inc.

*Norton* Biological reactor packing media by Aeration Engineering Resources Corp.

*Notim*™ Organic iron and tannin removal media by Matt-Son, Inc.

*NoVOCs*™ In-well stripping process for volatile organic compound removal by EG&G Environmental.

*Novus*® Emulsion polymers by Betz.

*No-Wear*™  Traveling bridge filter backwash shoe by Davis Water & Waste Industries, Inc.

*No-Well*  Pier-mounted traveling water screen design that does not require channel-type intake by FMC Corp., MHS Division.

**NOx**  See "nitrogen oxides."

*Noxidizer*™  Incineration system by John Zink Co.

*Noxon*  Decanter centrifuge by Purac Engineering, Inc.

*NOxOut*  Nitrogen oxide reduction system by Nalco Chemical Co.

*Nozzle Air*  Dissolved-air flotation aeration system by Envirotech Company.

**NPDES**  See "National Pollutant Discharge Elimination System."

**NPE**  Nonyl phenol ethoxylates.

**NPL**  See "National Priorities List."

**NPSH**  See "net positive suction head."

**NPSHA**  Net positive suction head available.

**NPSHR**  Net positive suction head required.

**NPT**  National pipe thread.

**NRA**  National Rivers Authority.

**NRWA**  National Rural Water Association.

**NSF**  National Science Foundation.

**NSPS**  See "New Source Performance Standards."

**NSSC**  Neutral sulfite semichemical pulping process.

**NTIS**  National Technical Information Service.

*N-Trak*  Test kit to determine nitrogen content of water by Hach Company.

**NTU**  See "nephelometric turbidity unit."

**nuclear magnetic resonance (NMR)**  An analytical technique used to detect and distinguish between nuclear particles in a sample using magnetic fields.

*Nuclepore*®  Membrane cartridge filter by Costar.

**NUG** Nonutility generator.

*Nu-Notch Mushroom* Cast iron air diffuser by the former Knowles Mushroom Ventilator Co.

*NuTralite®* Odor control product for neutralizing disulfide and other odors by NuTech Environmental Corp.

*Nu-Treat* Flocculator/clarifier by Envirex, Inc.

**nutrient** Any substance that is assimilated by organisms to promote or facilitate their growth.

*Nutrigest®* Clarifier by Smith & Loveless, Inc.

*NVCU™* Vapor control unit by NAO Inc.

*N-Viro* Pasteurization and chemical fixation process to disinfect and stabilize sludge by N-Viro Energy Systems, Inc.

**NWPA** Nuclear Waste Policy Act of 1982.

**NWSIA** National Water Supply Improvement Association. Former name of "American Desalting Association."

**nylon** Plastic compound that offers excellent load bearing capability, low frictional properties, and good chemical resistance.

# O

**obligate aerobes**   Bacteria that can survive only in the presence of dissolved oxygen.

**obligate anaerobes**   Bacteria that can survive only in the absence of dissolved oxygen.

*OBS®*   Turbidity sensors by D&A Instrument Company.

**occlusion**   An absorption process where one solid material adheres to another, sometimes resulting in coprecipitation.

**ocean disposal**   The discharge or disposal of wastes or sludges in ocean water.

**ocean dumping**   Disposal of wastes in the ocean or seas.

**Ocean Dumping Act (ODA)**   Authorizes regulation of intentional ocean disposal of materials, related research, and the establishment of marine sanctuaries.

**ocean incineration**   The burning of wastes on ocean-going vessels, in waters remote from land.

**OCPSF**   Organic chemicals, plastics, and synthetic fibers.

**OD**   Outside diameter.

**ODA**   See "Ocean Dumping Act."

*Odin*   Packaged water treatment plant by Davis Water & Waste Industries, Inc.

*Odophos*   System for controlling hydrogen sulfide and removing phosphorus by Davis Water & Waste Industries, Inc.

*Odor Buster®*   Aeration system used to reduce odors at pump stations and plant headworks by United Industries, Inc.

**Odorgard™** Enhanced hypochlorite air scrubbing process by ICI Katalco Co.

**OdorLok™** Hydrogen sulfide corrosion and odor control system by Eaglebrook, Inc.

**OdorMaster™** Electrolytic gas scrubber type odor control system by Pepcon Systems, Inc.

**Odor-Ox** Multistage dry chemical air scrubber by Purafil, Inc.

**odor threshold** See "threshold odor number."

**OEM** Original equipment manufacturer.

**offal** Trimmings and viscera of butchered animals.

**off-gas** The gaseous emissions from a process or equipment.

**offset** The requirement for a proposed air pollutant generator to reduce emissions or obtain emission reductions from other facilities to compensate for new emissions.

**OFR** See "overflow rate."

**OGWDW** U.S. Office of Groundwater and Drinking Water.

**OHL** Overhung load.

**ohm** Unit of electrical resistance where a potential difference of one volt produces a current of one ampere.

**Oil Grabber** Oil skimming system by Abanaki Corp.

**OilMaster** Oil/water separator by National Fluid Separators, Inc.

**Oil-Minder** Submersible pump unit by Stancor, Inc.

**Oil Pollution Act (OPA)** 1990 U.S. federal law that places liability on tank owners or operators for removal costs and damages if oil or other hazardous materials are spilled or discharged.

**oils and grease** Common term used to include fats, oils, waxes, and related constituents found in wastewater.

**oil skimmer** A device used to remove oil from a water's surface.

*Oilspin II*   Hydrocyclone by Serck Baker, Inc.

**oily wastewater**   An oil-in-water emulsion in which oil is dispersed in the water phase.

*OKI*   Mechanical aerator by Outomec USA, Inc.

**old growth forest**   A forest with a large percentage of old trees that have never been cut, or have not been cut for many years.

*Oleofilter*™   Filter for removal of hydrocarbons from water by Aprotek, Inc.

**olfactometer**   A device used by a panel of testers to compare the odor from an ambient air sample to reference samples of varying dilutions to determine odor strength.

**oligotroph**   Bacteria that grow in a medium containing less than 1.0 mg/L organic carbon.

**oligotrophic lake**   A deep lake deficient in organic materials whose waters contain a high degree of dissolved oxygen and low biochemical oxygen demand.

**O&M**   Operation and maintenance.

*Omega*   Horizontal rotor aerator by Purestream, Inc.

*Omega®*   Lime slaker and feeder package by Merrick Industries, Inc.

*O2 Minimizer®*   Process controller used to control oxygenation of mixed liquor by Schreiber Corp.

*Omnichlor*   Sodium hypochlorite generator for marine applications by Eltech International Corp.

*Omniflo®*   Sequencing batch reactor wastewater treatment system by Jet Tech, Inc.

*Omnipure*   Marine sewage treatment plant product line by Eltech International Corp.

**oncogenic**   A chemical or agent with tumor causing properties.

*Onguard®*   Corrosion monitor by Ashland Chemical, Drew Division.

**oocyst** An outer shell that protects an organism in the environment.

**OPA** See "Oil Pollution Act."

**opacity** The degree to which emissions reduce the transmission of a beam of light, expressed as a percentage of the light that fails to penetrate a plume of smoke.

**open burning** The combustion of solid waste without (a) containment of combustion reaction in an enclosed device, (b) control of the emission of the combustion products, or (c) controlling combustion air to maintain temperature for efficient burning.

**open channel** A natural or artificial channel in which fluid flows with a free surface, open to the atmosphere.

**open cycle cooling system** A system in which cooling water is discharged to a receiving body of water without being recycled.

**open dump** A land disposal site where solids are disposed of in a manner that does not protect the environment and is susceptible to open burning, as well as exposure to the elements, insects, and scavengers.

**opportunistic pathogen** A microbe that can cause disease in ill, very young, or elderly persons, but usually not in healthy individuals.

*Opti-Core* Polyvinyl chloride biological filter media formerly offered by B.F. Goodrich Co.

*Optimem*™ Reverse osmosis membranes by NWW Acumem, Inc.

*Optimer*™ Flocculant by Nalco Chemical Co.

*Optimum* Direct filtration water treatment plant by BCA Industrial Controls.

*Orbal* Oxidation ditch wastewater treatment system by Envirex, Inc.

**ORE** Orbital rod evaporation.

*Orec*   Ozone generating systems by Ozone Research & Equipment Corp.

*Organagro*®   Agricultural compost by Bedminster Bio-conversion Corp.

**organic loading**   The amount of organic matter applied to a treatment process.

**organic matter**   Substances containing carbon compounds, usually of animal or vegetable origin.

**organic nitrogen**   Nitrogen that is bound to carbon-containing compounds.

**organic phosphorus**   Phosphorus that is bound to carbon-containing compounds.

**organoclays**   A bentonite clay modified with quarternary amine, then granulated and blended with anthracite for use as an adsorbent.

**organophosphates**   Commonly used phosphorus-based organic pesticides that are relatively nonpersistent in the environment.

*Ori-Cast*   Cast elastomer material used in nonmetallic rectangular clarifier products by Oritex Corp.

**orifice plate**   (1) Flow measurement device that indicates flow as a function of differential pressure across a flow-restricting orifice. (2) Flow limiting device.

*Ori-Plastic*   Plastic material used in nonmetallic rectangular clarifier products by Oritex Corp.

**ORNL**   Oak Ridge National Laboratory.

**ORP**   See "oxidation-reduction potential."

*OSEC*™   Electrolytic chlorination system by Wallace & Tiernan, Inc.

**OSHA**   Occupational Safety and Health Administration. U.S. agency responsible for overseeing workplace health and safety.

*Osmo*®   Water purification systems by Osmonics, Inc.

**osmosis**  Movement of water from a dilute solution to a more concentrated solution through a permeable membrane separating the two solutions.

*Osmostill*  Distillation unit by Osmonics, Inc.

**osmotic pressure**  Excess pressure that must be applied to a concentrated solution to produce equilibrium and prevent the movement of a more dilute solution, through a semipermeable membrane, into the more concentrated solution.

**OST**  Office of Science and Technology. A U.S. EPA office.

**OSW**  Office of Saline Water.

*OTA® Aerator*  Rotor aerator by Scoti-Zahner, Inc.

**OTE**  Orbital tube evaporation.

**OUR**  Oxygen uptake rate.

**outfall**  The location where a storm or sanitary sewer is discharged into a receiving water body.

**outhouse**  See "privy."

**overburden**  The soil and rock overlying a mineral deposit that must be removed prior to the start of strip mining.

**overflow rate (OFR)**  An expression used to indicate the upward water velocity in a sedimentation tank, expressed as flow per day per unit of basin surface area. Also called "surface loading rate."

**overflow weir**  A weir over which excess water or wastewater is allowed to flow.

*Owamat®*  Oil/water separator by BEKO Condensate Systems Corp.

*Oxidair*™  Thermal oxidizer for soil remediation and off-gas treatment by EPG Companies, Inc.

**oxidant**  A chemical substance, such as chlorine or ozone, that is capable of promoting oxidation.

**oxidation**  (1) A chemical reaction in which an element or ion loses electrons. (2) The biological or chemical conversion of organic matter into simpler, more stable forms.

**oxidation ditch**  An extended aeration waste treatment process that occurs in an oval-shaped channel or ditch (also called a "race track"), with aeration provided by a mechanical brush aerator.

**oxidation pond**  An earthen wastewater basin in which biological oxidation of organic matter occurs naturally, or with assistance of mechanical oxygen transfer equipment.

**oxidation-reduction potential (ORP)**  The potential required to transfer electrons from an oxidant to a reductant, which indicates the relative strength potential of an oxidation-reduction reaction.

*Oxidator*  Combination aeration, flocculation, and sedimentation unit by Eimco Process Equipment Co.

**oxide**  A compound of an element with oxygen alone.

**oxidizing agent**  Any substance that can contribute electrons to a reaction.

*Oxifree®*  Ultraviolet disinfection system by Capital Controls Co., Inc.

*Oxigest®*  Cylindrical package extended aeration waste treatment plant by Smith & Loveless, Inc.

*Oxigritter*  Primary sewage treatment unit by Eimco Process Equipment Co.

*Oxitech®*  Resin conditioning process for total organic carbon reduction by U.S. Filter Corp.

*Oxitrace™*  Oxidant analyzer and monitor by Capital Controls Co., Inc.

*Oxitron™*  Fixed-film wastewater treatment plant by Krüger, Inc.

*Oxycat*  Air pollution abatement catalyst by Met-Pro Corp.

*OxyCharger*  Static aerator by Parkson Corp.

*Oxychlor*   Chlorine dioxide generator by International Dioxide, Inc.

*Oxyditch*   Oxidation ditch treatment system formerly offered by Chemineer-Kenics.

*Oxy Flo*   Mechanical aerator by Aqua-Aerobic Systems, Inc.

*Oxy-Gard*   Aeration control system that monitors dissolved oxygen level to control blower operation by Lamson Corp.

**oxygen**   A chemical element that comprises approximately 20% of the earth's atmosphere and is essential for biological oxidation.

**oxygen deficiency**   The additional amount of oxygen required to satisfy the oxygen requirement of wastewater.

**oxygen sag**   The temporary decrease in the dissolved oxygen level in a stream or river that occurs downstream from a point source of pollution.

**oxygen scavenger**   A chemical used to supplement mechanical deaeration.

**oxygen transfer**   The exchange of oxygen between a gaseous and a liquid phase.

**oxygen transfer rate**   The mass of oxygen transferred per unit time.

**oxygen uptake**   The amount of oxygen used during biochemical oxidation.

*Oxygun™*   Subsurface self-aspirating aerator by Framco Environmental Technologies.

*Oxyrapid*   Air diffusion and recycling system for activated sludge system by Infilco Degremont, Inc.

*Oxyrotor*   Surface brush aerator by Euroquip Fabrication, Ltd.

*Oxytrace™*   Chlorine residual analyzer by Capital Controls Co., Inc.

*Oxytrap*™    Wastewater aerator by DAS International, Inc.

*OZ*    Ozone generation equipment by Ozone Pure Water, Inc.

*Ozat*    Compact ozone generator by Ozonia North America.

**ozonation**    The process of using ozone in water or wastewater treatment for oxidation, disinfection, or odor control.

**ozonator**    An ozone generator.

**ozone**    A strong oxidizing agent with disinfection properties similar to chlorine; also used in odor control and sludge processing. Chemical formula is $O_3$.

**ozone generator**    Device used to produce ozone by passing air/oxygen through an electric field.

**ozone layer**    The portion of the stratosphere, extending from an altitude of approximately 20–50 km, in which naturally occurring ozone protects life on earth by filtering out harmful ultraviolet radiation from the sun.

*Ozonmat*®    Ozone analyzer by Z Polymetron.

# P

**PAC**  See "powdered activated carbon."

*PAC*  Valve positioner and controller by F.B. Leopold Co.

*Pace®*  Oil/water separator by Scienco/Fast Systems.

*Pacer*  Package water treatment plant by Roberts Filter Manufacturing Co.

*Pacesetter*  Liquid/liquid gravity separator by Envirotech Company.

*Pacific Flush Tank*  Former digestion equipment manufacturer acquired by Envirex, Inc.

**package plant**  Factory-assembled treatment plant generally incorporated in a single tank, or at most, several tanks.

**packed column**  A vertical vessel filled with packing material usually used to strip gases or degasify liquids.

**packing**  The fill material in a fixed-film reactor or stripping vessel that provides large surface area per unit volume.

*Pacpuri®*  Sodium hypochlorite generation system by Electrocatalytic, Ltd.

*PACT®*  Powdered activated carbon wastewater treatment process by Zimpro Environmental, Inc.

*Paddle Dryer*  Sludge dryer by Komline-Sanderson.

**paddle flocculator**  A flocculation device utilizing rotating baffles to accomplish mixing.

**PAH**  See "polycyclic aromatic hydrocarbons."

**paint filter test**  Test to determine free water content of sludge by noting whether water drains from a sample placed on filter paper.

*PakTOR*   Multicell packed bed reactor by General Filter Co.

**P-alkalinity**   See "phenolphthalein alkalinity."

**Palmer-Bowlus flume**   A portable, venturi-type flume used to measure water or wastewater flow.

**PAN**   See "peroxyacetyl nitrate."

**pandemic**   A worldwide epidemic.

*Para Cone*   Internally fed rotary fine screen by Andritz-Ruthner, Inc. (Western Hemisphere) and Contra-Shear Engineering, Ltd.

*Paraflash*   Forced circulation evaporator by APV Crepaco, Inc.

*Paraflow*   Plate heat exchanger by APV Crepaco, Inc.

**paraquat**   An herbicide resistant to microbial degradation that has been used to control marijuana; exposure to it can result in serious health effects or death.

**parasite**   An organism that lives either on or inside a larger host organism; the presence of the parasite is usually harmful to the host organism.

**parasitic bacteria**   Bacteria that require a living host organism.

*Para-Stat*   Static screen by Dontech, Inc.

*Paravap*   High-solids evaporator by APV Crepaco, Inc.

**parenteral solution**   A solution introduced into the body by a vein, muscle, or pathway other than the mouth.

*Parkwood*   Sewage treatment equipment product line by Longwood Engineering Co., Ltd.

**Parshall flume**   A fixed, venturi-type flume used to measure water or wastewater flow.

**parthenogenic**   Capable of reproduction by means of an unfertilized egg.

**partial closure**   The closure of a hazardous waste management unit at a facility that contains other active hazardous waste management units.

**partial pressure**  The pressure exerted by each gas in a mixture that is proportional to the amount of that gas in the mixture.

**particle counter**  Instrument used to measure the size and count the number of particles in water.

**particle size analysis**  Determination of the amounts of different particle sizes in a sample.

**particulate**  Usually considered to be a solid particle larger than 1 micron, or large enough to be removed by filtration.

**particulate organic carbon (POC)**  The portion of organic matter that can be removed by filtration through a 0.45-micron filter.

**particulate organic matter (POM)**  Material of plant or animal origin that is suspended in water, and usually capable of being removed by filtration.

*Partisol®*  Air sampler by Rupprecht & Patashnick Co.

**parts per million (ppm)**  A common unit of measure used to express the number of parts of a substance contained within a million parts of a liquid, solid, or gas. Generally interchangeable with "milligrams per liter" in dilute solutions and water treatment calculations.

*PASS®*  Polyaluminum-sulfate coagulant by the Alumina Company Ltd.

*Passavant*  Equipment product line by Zimpro Environmental, Inc.

**passivation**  The changing of a chemically active surface of a metal to a much less reactive state. Usually done to stainless steel by immersion in an acid bath.

**passive screen**  Intake screening device that does not employ mechanical cleaning.

**pasteurization**  A process for killing pathogenic organisms by applying heat for a specific period of time.

**pathogen**   Highly infectious, disease producing microbes commonly found in sanitary wastewater.

*Pathwinder®*   Screenings conveyor by Serpentix Conveyor Corp.

*Patriot*   Fluid recovery treatment system for spent coolant and oils by SanTech, Inc.

*Paygro*   In-vessel composting system by Compost Systems Co.

**PC**   See "physical-chemical treatment."

**PCB**   See "polychlorinated biphenyl."

*PCC*   Process Combustion Corporation.

**PCE**   Perchloroethylene.

*PCI*   Patterson Candy International, Ltd.

*PCI*   Pollution Control, Inc.

**pCi**   See "picocurie."

**PCV**   See "positive crankcase ventilation."

**PD**   Positive displacement.

*PDC™*   Polymer dosage control system by Andritz-Ruthner, Inc.

*PDM/Roediger®*   Anaerobic sludge digestion system by Pitt-Des Moines, Inc.

**PE**   Professional engineer.

*Peabody Floway*   Former name of Floway Pumps, Inc.

*Peabody Welles*   Former manufacturer whose product lines were acquired by Aerators, Inc.

**peak flow**   Excessive flows experienced during hours of high demand, usually determined to be the highest 2-hr flow expected to be encountered under any operational conditions.

**peaking factor**   The ratio of peak to average flow.

*Pearlcomb®*   Fine-bubble diffuser by FMC Corp., MHS Division.

*Pearth*   Anaerobic digester gas mixing system by Envirex, Inc.

**peat** Material formed by partial decay of marsh vegetation with a moisture content greater than 75%.

*PEI* Purac Engineering, Inc.

**PEL** See "permissible exposure limit."

*Pelican* Wall-mounted clarifier scum removal system by Copa Group.

*Pelldry* Liquid-absorbing pellet by the former Sheldahl Industrial Absorbents.

*Pelletech®* Indirect sludge dryer and pelletizing unit by Wheelabrator Clean Water Systems, Inc., Bio Gro Division.

**Pelton wheel** An impulse hydraulic turbine that may be used an energy recovery device in high head applications such as seawater reverse osmosis.

**PEMS** Predictive emissions monitoring systems.

*Penberthy* Former manufacturer whose product line is now offered by Chemineer, Inc.

*Penfield®* Water treatment product line by U.S. Filter Corp.

*Penro* Reverse osmosis systems by Penfield Liquid Treatment Systems.

**penstock** A pipe that transports water to a turbine for the production of hydroelectric energy.

*PentaPure®* Disinfecting resin by WTC Industries, Inc.

*Pentech* Former manufacturer whose product line is now offered by Chemineer, Inc.

**PERC** See "perchloroethylene."

**perchloroethylene (PERC)** An organic solvent used in dry cleaning to remove dirt from clothing.

*Percol* Polyelectrolyte used to enhance liquid/solid separation by Allied Colloids, Inc.

**percolating filter** See "trickling filter."

**percolation** The flow or trickling of a liquid downward through a contact or filtering medium.

**percolation test** Test used to determine the water-absorbing capacity of soil where the drop in water level in a test hole is measured over a fixed time period.

*Perc-Rite®* Filtration system by Waste Water Systems, Inc.

**perc test** See "percolation test."

**performance ratio** A unit of measurement used to characterize evaporator performance, expressed as the mass of distillate produced per unit of energy consumed.

**peristaltic pump** A type of progressing cavity pump where the fluid is squeezed through a flow tube by external rollers.

*Perma-buoy* Foam-filled fiberglass flight for chain and flight sludge collector by Jeffrey Division/Indresco.

*PermaCare* Chemical additives for use with reverse osmosis systems by Houseman Ltd.

*Permachem* Packages containing chemical compositions and reagents by Hach Company.

**permafrost** Permanently frozen subsurface soil layer in the polar regions.

*Permaglas®* Storage tank coating system by A.O. Smith Harvestore Products, Inc.

*Permaklip* Belt seam for filter press belt by Tetko, Inc.

*Permalife* Chain and flight sludge collector components by Jeffrey Division/Indresco.

**permanent hardness** Hardness associated with sulfates, chlorides, and nitrates of calcium and magnesium that remain after boiling.

**permanent threshold shift (PTS)** A permanent hearing loss for a certain sound frequency.

*Permasep®* Reverse osmosis products by E.I. DuPont de Nemours, Inc.

**permeability** The property of a filter medium to permit a fluid to pass through it under the influence of pressure.

**permeate** The liquid that passes through a membrane.

**permeator** A pressure vessel containing semipermeable membranes.

*PermeOx®* Solid peroxygen by FMC Corp., Peroxygen Chemicals Division.

**permissible exposure limit (PEL)** Workplace exposure limit for 600 industrial chemicals, established by the Occupational Safety and Health Administration.

*Permofilter* Horizontal multiple-cell pressure filter by U.S. Filter Corp.

*Permupak* Package water treatment plant by U.S. Filter Corp.

*PermuRO* Reverse osmosis system by U.S. Filter Corp.

*Permutit®* Water treatment product line by U.S. Filter Corp.

**peroxone** A blend of ozone and hydrogen peroxide used for disinfection and odor control.

*Perox-Pure®* UV-catalyzed hydrogen peroxide system by Peroxidation Systems, Inc.

*Perox-serv™* Odor control services by Peroxidation Systems, Inc.

*Perox-stor™* Hydrogen peroxide user service by Vulcan Peroxidation Systems, Inc.

**peroxyacetyl nitrate (PAN)** A secondary pollutant and major component of photochemical smog formed when reactive hydrocarbons and oxides of nitrogen combine in the presence of sunlight.

*Perpac* Surface water treatment plant by Vulcan Industries, Inc.

**PERT** Program evaluation review technique.

**pervaporation (PV)** A process in which membranes are used to remove volatile organic compounds from an aqueous stream.

**pesticides**  A chemical used to kill undesired insects or animals.

*PET*™  Demineralizer system by U.S. Filter Corp.

**petri dish**  A covered dish containing an agar media used in the laboratory to cultivate bacteria.

**petrochemicals**  Products or compounds produced by the processing of petroleum and natural gas hydrocarbons.

*Petro-Flex*®  Portable holding tank by Aero Tec Laboratories, Inc.

**petroleum**  The crude oil removed from the earth and the oils derived from tar sands, shale, and coal.

*Petrolux*  Ceramic membrane filter by U.S. Filter Corp.

*Petro-Pak*  Coalescing media for oil removal system by McTigue Industries, Inc.

*Petro-Xtractor*™  Water well oil skimmer by Abanaki Corp.

**pezodialysis**  Membrane process to remove salt from solution where salt, rather than water, passes through the membrane.

**PFBC**  Pressurized fluidized bed combustion.

**PFD**  Process flow diagram.

**PFO**  Power fail open.

**PFRP**  Process to further reduce pathogens.

*PFT*  Former equipment manufacturer acquired by Envirex, Inc.

**pH**  The reciprocal of the logarithm of the hydrogen ion concentration in gram moles per liter.

**pharmaceutical-grade water**  See "USP-purified water."

**phase**  The state of a substance: solid, liquid, or vapor.

**PHD**  Peak hourly demand.

*PhD²*  Portable multiple gas detector by Biosystems, Inc.

**phenolphthalein**  A color indicator that changes from colorless to pink/red, used to measure alkalinity.

**phenolphthalein alkalinity**   Alkalinity determined by titration with sulfuric acid to pH 8.3, indicated by a color change of phenolphthalein, and expressed as mg/L of calcium carbonate.

**phenols**   Organic pollutant also known as carbolic acid occurring in industrial wastes from petroleum processing and coal coking operations.

*Phoenix Press*   Belt filter press by Phoenix Process Equipment Co.

**phosphate**   A salt or ester of phosphoric acid.

**phosphorus**   A nutrient that is an essential element of all life forms.

*Phostrip*   Biological system for phosphorus and biochemical oxygen demand removal by TETRA Technologies.

*Phosver*   Reagent chemical used to determine phosphate concentration in water by Hach Company.

**photic zone**   The upper level of a water body into which light penetrates.

**photochemical smog**   A form of air pollution that results in an atmospheric haze and is caused by the reaction of sunlight with volatile organic compounds, nitrogen oxides, and other pollutants produced by combustion processes.

**photosynthesis**   The process of converting carbon dioxide and water to carbohydrates, activated by sunlight in the presence of chlorophyll.

**phototrophs**   Organisms that rely on the sun for energy.

**photovoltaic cell**   A device that utilizes crystalline materials to convert light from solar radiation directly into electricity.

*pHREEdom*™   Cooling water treatment chemicals by Calgon Corp.

**phys-chem**   See "physical-chemical treatment."

**physical-chemical treatment**   Treatment processes that are nonbiological in nature.

**physical treatment**   A water or wastewater treatment process that utilizes only physical methods such as filtration or sedimentation.

**phytoplankton**   Algae that exist floating or suspended freely in a body of water.

*Picabiol®*   Activated carbon purification process for potable water by Pica.

**pickets**   Vertical paddles used in a gravity thickener.

**pickle liquor**   Waste acid from steel pickling process.

**pickling**   A chemical or electrochemical method of removing mill scale and rust from steel by washing or immersing in an acid or salt solution.

**picocurie (pCi)**   A unit of radioactivity equal to $3.7 \times 10^{-12}$ curie.

**PICs**   See "products of incomplete combustion."

**P&ID**   Process and instrumentation diagram.

**piezometer**   An instrument, consisting of a small pipe and manometer, fitted to the wall of a pipe or container to measure pressure head.

**piezometric head**   The elevation plus pressure head.

**pig**   A water-propelled internal pipe cleaner.

**pigging**   A pipeline cleaning procedure using water-propelled pigs to scour solids from the interior walls of a pipe.

*Pilgrim*   Former equipment manufacturer acquired by Andritz-Ruthner, Inc.

**piling**   Timbers, concrete posts, or other structural elements that are embedded into the ground to support a load.

**pilot plant**   A water or wastewater treatment plant that is smaller than full scale used to test and evaluate a treatment process.

**Pinch Press**  High-pressure filtration and dewatering device by Waste Tech, Inc.

**pinch valve**  A valve where sealing is achieved by one or more flexible elements that can be pinched to stop flow.

**pin floc**  Small floc particle.

**pintle chain**  Chain extensively used for elevating and conveying, consisting of one-piece links cast with two offset sidebars and coupled with steel pins.

**pipe gallery**  A passageway to provide access for installation and maintenance of underground pipes and valves.

*Pipeliner*  In-line grinder/cutter device by Robbins Myers, Inc.

**pipe spool**  A prefabricated section of piping.

**pipet**  A calibrated glass tube used to deliver prescribed volumes of liquids, usually less than 10 mL.

*Pista® Grit*  Vortex-type grit removal system by Smith & Loveless, Inc.

**piston pump**  A reciprocating pump whose piston normally incorporates a sliding seal with the cylinder wall.

**pitch**  (1) The length of one link of chain measured from pin centerline to pin centerline. (2) The distance between the centers of adjacent tubes.

*Pit Hog®*  Sludge pumping system by LWT, Inc.

**pitot tube**  Flow measurement device that measures the velocity head of a liquid stream as the difference between the static head and the total head.

*Pittchlor*  High-test calcium hypochlorite product by PPG Industries.

**pitting**  Localized corrosion causing attacks over small surface areas that may reach considerable depths.

**PIV**  Positive infinitely variable.

*Planet*  Rotary distributor for fixed-film reactor by Simon-Hartley, Ltd.

**plankton**   Small, passively floating or weakly swimming animal and plant life of a body of water.

*PLASdek*   Cooling tower fill by Munters.

**plate-and-frame press**   A batch process sludge dewatering device in which sludge is pumped through a series of parallel plates fitted with filter cloth.

**plate count**   The number of microbe colonies that develop on a laboratory test dish after a fixed incubation period.

*Plate-Pak®*   Vane mist eliminators by ACS Industries, Inc.

**plate settler**   Clarifier with enhanced sedimentation through the use of steeply inclined plates.

*Platetube*   Porous diffuser plates by Walker Process Equipment Co.

**PLC**   Programmable logic controller.

**plug flow**   Flow conditions where fluid and fluid particles pass through a tank and are discharged in the same sequence that they enter.

**plugging factor**   See "silt density index."

**plume**   The measurable or visible impact of a discharge into the air or a body of water.

**plunger pump**   A reciprocating pump whose plunger does not contact the cylinder walls, but enters and withdraws from it through packing glands.

*Plus 5*   Air diffuser by EnviroQuip International Corp.

*Plus 150™*   Laboratory water system by U.S. Filter Corp.

**PM**   Preventative maintenance.

**PM$_{10}$**   Airborne particulate matter having a diameter equal to or smaller than 10 microns.

*PMD™*   Pipe-mounted diffuser by Environmental Dynamics, Inc.

**PNA**   Polynuclear aromatics.

**pneumoconiosis**   Lung disease that results from chronic exposure to dusts.

**POC** Particles of complete combustion.

*Pocket Pal* Portable pH tester by Hach Co.

**POCs** See "particulate organic carbon."

**POE** See "point-of-entry."

**pOH** The negative logarithm of the hydroxyl ion concentration approximated by the relationship: $14 - pH = pOH$.

**POHC** See "principal organic hazardous constituent."

**point-of-entry (POE)** Location of a water treatment or water quality control device at the point drinking water enters a house or building for the purpose of reducing the contaminants in the drinking water distributed throughout the house or building.

**point-of-use (POU)** Location of a water treatment or water quality control device at a faucet in an individual household.

**point source discharge** A pipe, ditch, channel, or other container from which pollutants may be discharged.

*Pol-E-Z®* Emulsion polymer to enhance solids/liquid separation by Calgon Corp.

**pollutant** A substance, organism, or energy form present in amounts that impair or threaten an ecosystem to the extent that its current or future uses are precluded.

**pollution** The presence of a pollutant in the environment.

*Pollutrol* Former equipment manufacturer.

*Polly Pig* Internal pipeline cleaner by Knapp Polly Pig, Inc.

*Polyad™* Fluidized bed volatile organic compound emission control system by Weatherly, Inc.

*Poly-Alum* Polymerized inorganic coagulant by Rochester Midland Co.

*PolyBlend* Polymer mixing/feeding products by Stranco, Inc.

*Poly Boss*   Instrument to check settling velocity in clarifier feedwell by WesTech Engineering, Inc.

*Polybrake*   Cleaning product for removal of polymers by AquaPro, Inc.

**polychaete worm**   A small worm common in seas and estuaries, and often chosen for bioassays of coastal regions.

*Polychem*   Molded plastic products for rectangular clarifiers by Budd Co.

**polychlorinated biphenyl (PCB)**   Class of hazardous organic compounds considered probable carcinogens, formerly used in the manufacture of electrical insulation and heating and cooling equipment.

**polycyclic aromatic hydrocarbons (PAH)**   A group of aromatic compounds, many of which are carcinogenic, formed by industrial processes and during the burning of gasoline, coal, and other substances.

**polyelectrolyte**   A compound consisting of a chain of organic molecules used as coagulants or coagulant aids. See also "polymer."

**polyethylene**   An inexpensive plastic with a low coefficient of friction and excellent abrasion, impact, and chemical resistance.

*Poly-Filter*   Plate and frame filter press by the former Clow Corp.

*Polyjet*   Flow control valve by Bailey.

*Poly-Links*   Nonmetallic sludge collector chain by NRG, Inc.

*Polymair*   Package polymer processing system by Acrison, Inc.

**polymaleic acid**   A high-performance scale control additive.

*Polymaster®*   Polymer mixing/feeding system by Komax, Inc.

*PolyMax*   Polymer mixing/feeding products by Semblex, Inc.

**polymer**   (1) High-molecular-weight compounds derived by the recurring addition of similar molecules. (2) Common term for "polyelectrolyte."

*Polymer Piping & Materials*   Former equipment manufacturer whose products are now offered by Jaeger Products, Inc.

*PolyMixer*   Polymer mixing/feeding system by Atlantes Chemical Systems, Inc.

*Polymizer*   Centrifuge by Alfa Laval Separation, Inc.

**polynuclear aromatic hydrocarbons**   See "polycyclic aromatic hydrocarbons."

*Polyozone*   Ozone equipment product line by Polymetrics.

*Polypak*   Polymer feeding system by Leeds & Northrup.

**polyphosphates**   Phosphate compounds used as sequestration agents to prevent formation of iron, manganese, and calcium carbonate deposits.

*PolyPress*   Plate and frame filter press by Star Systems.

*PolyPro*   Dry polymer feed system by AquaPro, Inc.

*Polyseed®*   Bacterial culture for biochemical oxygen demand seeding by Polybac Corp.

*Poly-Stage™*   Modular air scrubber system by Davis Industries, Process Division.

*PolyThickener*   Waste activated sludge thickener by Walker Process Equipment Co.

*Polytox™*   Biological toxicity test kit by Polybac Corp.

*PolyTube*   Tubular air diffuser formerly offered by Walker Process Equipment Co.

**polyurethane**   An elastomer with tensile strength and abrasion resistance greater than that of natural rubber, and capable of being formulated for injection molding or casting.

**polyvinyl chloride (PVC)**   A thermoplastic with excellent corrosion resistance that is widely used to manufacture pipe and biological filter media.

**POM**   See "particulate organic matter."

**ponding**   See "pooling."

*Pond-X®*   Odor control product by NuTech Environmental Corp.

*Pontoon*   Floating cover for anaerobic digesters by FMC Corp., MHS Division.

**pooling**   The formation of pools of liquid on the surface of a clogged filter.

*Porcupine*   Indirect contact sludge dryer by the Bethlehem Corp.

*Poro-Carbon*   Automatic liquid filter system by R.P. Adams Co., Inc.

*Poro-Edge*   Automatic water strainer by R.P. Adams Co., Inc.

**porosity**   The ratio of void volume to total bulk volume.

*Poro-Stone*   Automatic liquid filter system by R.P. Adams Co., Inc.

**porous disk diffuser**   A circular, fine-bubble aeration device of porous plastic or ceramic construction.

*Port-A-Berm™*   Portable secondary containment system by Aero Tec Laboratories, Inc.

*Portacel®*   Gas chlorination system by F.B. Leopold Co., Inc.

*Porta-Cleanse*   Submersible mixer for wet well pump station by ITT Flygt Corp.

*Porta-Feed®*   Chemical handling system by Nalco Chemical Co.

*Porta-Tank*   Liquid storage system by Environetics, Inc.

**Portland cement**   Cement made by heating a slurry of crushed chalk or limestone and clay to clinker in a kiln, before grinding and adding gypsum.

**Posi-Clean** Wedgewire straining element by Tate Andale, Inc.

**Posirake** Reciprocating rake bar screen by Zimpro Environmental, Inc.

**positive crankcase ventilation (PCV)** A method of reducing automobile engine emissions by directing crankcase emissions into the cylinders for combustion.

**positive displacement pump** A pump where liquid is drawn into a cavity and its pressure is increased, forcing the liquid through an outlet port into the discharge line.

**Positive Seal** Rotary distributor for trickling filter by Walker Process Equipment Co.

**postchlorination** Addition of chlorine after completion of other treatment processes.

**posttreatment** Treatment of finished water or wastewater to further enhance its quality.

**potable water** See "drinking water."

**potassium permanganate** A crystalline salt of potassium and manganese used for taste and odor control and iron/manganese oxidation. Chemical formula is $KMnO_4$. Also called "purple salt."

**potentially responsible party (PRP)** An individual or company identified as potentially liable for cleanup or payment for cost of cleanup of a hazardous waste site.

**POTW** See "publicly owned treatment works."

**POU** See "point-of-use."

**pounds per square inch, absolute (psia)** The total pressure in a system, which is the sum of the gauge pressure and atmospheric pressure, and expressed in terms of the pounds of force exerted per square inch.

**pounds per square inch, gauge (psig)** The pressure measured by a gauge and expressed in terms of the pounds of force exerted per square inch.

**POU/POE** Point-of-use/point-of-entry.

**powdered activated carbon (PAC)**   A powered form of activated carbon fed as a slurry to water to absorb organics, particularly taste- and odor-causing constituents.

*Powder Pop®*   Chlorine test dispenser by HF Scientific, Inc.

*Powdex*   Combination ion exchange and filtration unit by Graver Co.

*PoweRake*   Reciprocating rake bar screen by the former EnviroFab, Inc.

*Power Backwash™*   Concurrent air and water filter backwash system by TETRA Technologies, Inc.

*Power Brush™*   Internal pipeline cleaner by Pipeline Pigging Products, Inc.

*Powerhouse®*   Alkaline cleaner/degreaser by Technical Products, Corp.

*Powermatic*   Reciprocating rake bar screen by Brackett Green, Ltd.

*Power Mizer™*   Multistage centrifugal blower by Spencer Turbine Co.

*Power Units™*   Magnetic pipe treatment unit by Aqua Magnetics International, Inc.

*PoweRupp®*   Electrically driven diaphragm pump by Warren Rupp, Inc.

*Powrclean*   Front-cleaned multiple-rake bar screen by Aerators, Inc.

*Powr-Trols*   Custom engineered pump station motor control panel by Healy-Ruff Co.

*PO\*WW\*ER™*   Wastewater treatment process to reduce residual solids by ARI Technologies.

*Poz-O-Lite®*   A lightweight aggregate by Conversion Systems, Inc.

*Poz-O-Tec®*   Scrubber sludge and stabilizing additive mixture that produces a stable, nonleaching, monolithic mass of low permeability by Conversion Systems, Inc.

**pozzolonic**　Finely divided materials such as fly ash that aid in forming compounds possessing cementitious properties.

**ppb**　Parts per billion.

**ppm**　See "parts per million."

**ppmv**　Parts per million by volume.

*PQ®*　Sodium silicate corrosion inhibitor by PQ Corp.

*Praestol®*　Polymer products by Stockhausen, Inc.

**preaeration**　Aeration of wastewater prior to primary sedimentation to improve its treatability.

**prechlorination**　The application of chlorine before other treatment processes.

**precipitation**　The phenomenon that occurs when a substance held in solution passes out of solution into a solid form.

*Precipitator*™　Package treatment plants by U.S. Filter Corp.

*Precision*　Fine-bubble tube diffuser by FMC Corp., MHS Division.

**precoat filter**　Filter using a thin layer of very fine material, such as diatomaceous earth, to coat the filter surface before filtration cycles.

**precursor**　A substance or compound from which another substance or compound is formed; usually refers to an organic compound capable of being formed into a trihalomethane.

*PreFLEX®*　Skid-mounted pretreatment system for reverse osmosis and demineralizer systems by U.S. Filter Corp.

**preliminary treatment**　Treatment steps including comminution, screening, and grit removal that prepare wastewater influent for further treatment.

*Prerostal*　System of adjusting pumping volume to inflow rate by Envirotech Company.

**presedimentation** A pretreatment process used to remove sand, gravel, or other gritty material before subsequent treatment.

**pressate** The liquid waste stream from a filter press.

*PressMaster*™ Hydraulic sludge dewatering unit by Eimco Process Equipment Co.

**pressure filter** Filter unit enclosed in a vessel that may be operated under pressure.

*Pressveyor* Hydraulic screenings press/conveyor by Hycor Corp.

*Prestex* Filter belt for belt filter press by Tetko, Inc.

**pretreatment** (1) The initial water or wastewater treatment process that precedes primary treatment processes. (2) The treatment of industrial wastes to reduce or alter the characteristics of the pollutants prior to discharge to a publicly owned treatment works.

*Pretreat Plus*™ Antiscalant/dispersant by King Lee Technologies.

*Pretreat SDP*™ Antiscalant for small-capacity reverse osmosis systems by King Lee Technologies.

**prevention of significant deterioration (PSD)** A Clean Air Act regulatory program that establishes a minimum air quality baseline for particular pollutants in specific areas.

**primary clarifier** Sedimentation basin that precedes secondary wastewater treatment.

**primary contaminant** A drinking water contaminant with health-related effects.

**primary industry categories** The 34 types of industrial facilities that require the best available technology for toxic water pollutants under the Clean Water Act.

**primary MCL** EPA-mandated maximum contaminant level in drinking water based on health effects.

**primary pollutant**   A pollutant that exists in the environment in the same form as when it was released.

**primary sedimentation**   Principal form of primary wastewater treatment utilizing clarifiers to reduce the solids loading on subsequent treatment processes.

**primary sludge**   Sludge produced in a primary waste treatment unit.

**primary treatment**   Treatment steps including sedimentation and/or fine screening to produce an effluent suitable for biological treatment.

*Prima-Sep*®   Specialty clarifier with tray separator by Graver Co.

*Primox*®   Oxygen injection system for primary sewage by BOC Gases.

*Princeton Clearwater*   Former equipment manufacturer.

**principal organic hazardous constituent (POHC)**   Hazardous organic compounds that may form products of incomplete combustion when incinerated.

**priority pollutants**   A list of approximately 126 chemicals identified as toxic pollutants by the Clean Water Act.

**privately owned treatment works**   A treatment works that is not owned or operated by a state or municipality. Not a "POTW."

**privatization**   The involvement of nonpublic and entrepreneurial interests in project development, ownership, and/or operation of municipal facilities such as water and wastewater treatment systems.

**privy**   A pit toilet. Also called an "outhouse."

*Probiotics*™   Lagoon sludge oxidation products by Bio Huma Netics, Inc.

*ProBlend*   Liquid polymer feed system by AquaPro, Inc.

**process wastewater**   Wastewater generated during manufacture or production processes.

**process water**   Water that is used for, or comes in contact with, an endproduct or the materials used in an endproduct.

*Prochem®*   Agitator and mixer products by Chemics, Inc.

**produced water**   All waters produced during the production of oil and gas.

**products of incomplete combustion (PICs)**   Carbon monoxide, hydrocarbons, and other organic matter generated when organic materials are burned.

**product staging**   Reverse osmosis process configuration where the product from one stage is used as feedwater on a subsequent stage to improve product water quality.

**product water**   Water produced as a result of treatment.

*Profiler*   Sludge blanket level and suspended solids monitor by Mt. Fury Co., Inc.

**profile wire**   Term used to describe specially shaped wire that is generally triangular or trapezoidal in cross-section. Also called "wedgewire."

**progressing cavity pump**   Pump used for viscous fluids, including sludge, that consists of a single-threaded shaft rotor rotating in a double-threaded rubber stator.

**progressive cavity pump**   See "progressing cavity pump."

*ProGuard*   Backwashable cartridge filters by ProGuard Filtration Systems.

*Promal*   Pearlitic malleable iron chain material by FMC Corp., MHS Division.

*ProMix™*   Polymer wetting and blending unit by BlenTech Inc.

*ProPack*   Random trickling filter media by the former Gray Engineering Co.

*PRO\*PEL*   Bead-type air pollution control catalysts by Prototech Co.

**prophylaxis** The observance of procedures necessary to prevent disease.

**proportional weir** A weir whose discharge is directly proportional to the head.

*Proportioneer* Mixer by Lightnin.

*Propulsair* Aspirating aerator by Eimco Process Equipment Co.

*ProSep* Membrane treatment system for industrial wastes by Advanced Membrane Technology, Inc.

*Prosonic™* Ultrasonic level control system by Endress+Hauser.

*Prosser/Enpro* Pump product line by Crane Pumps & Systems.

*ProTec RO™* Powder antifoulant by King Lee Technologies.

**Protista** The class of living organisms that includes algae, bacteria, and protozoa.

*Protoc* Total organic carbon analyzers by Tytronics, Inc.

**protozoa** A group of microorganisms including amoebas, flagellates, and ciliates that feed on bacteria and other protists and reproduce by binary fission.

*PRO\*VOC* Bead-type air pollution control catalysts by Prototech Co.

**PRP** See "potentially responsible party."

**PSD** See "prevention of significant deterioration."

**pseudo-hardness** The action exhibited by sea, brackish, and other waters containing high concentrations of sodium that interferes with the normal behavior of soap.

**pseudomonas** A common rod-shaped aerobic bacteria.

**psi** Pounds per square inch.

**psia** See "pounds per square inch, absolute."

**psig** See "pounds per square inch, gauge."

**PSRP** Process to significantly reduce pathogens.

**Psychoda flies**   Small, dark-colored flies that creates a nuisance by breeding in trickling filter beds. Commonly known as "filter flies."

**psychrometer**   An instrument used to determine the relative humidity and vapor tension of the atmosphere.

**psychrophiles**   Bacteria that grow best at temperatures below 20°C.

**PTS**   See "permanent threshold shift."

**publicly owned treatment works (POTW)**   Treatment works owned by a state or municipality, including sewers, pipes, or other conveyances used to convey wastewater.

**public water system**   A system that pipes water for human consumption to at least 25 people or has 15 or more surface connections.

**PUC**   Public Utilities Commission.

**pug mill**   A grinder used to reduce the size of solid waste or sludge to facilitate further treatment or disposal.

*Pullman Power Products*   Chimney and storage silo product line by Wheelabrator Clean Air Systems, Inc.

*PullUp*   Removable aeration header and drop pipe assembly by Aerators, Inc.

**pulsation dampener**   A device using air or other compressible gas to absorb pressure irregularities and induce uniform flow in pump suction or discharge lines.

*Pulsator®*   Solids contact clarifier by Infilco Degremont, Inc.

*Pulsatron®*   Electronic metering pump by Pulsafeeder.

*Pulsemate™*   Metering pumps and controls by Wallace & Tiernan, Inc.

*Pulse Mix*   Short-term backwashing process to regenerate sand filter media by Zimpro Environmental, Inc.

**pump**   A mechanical device used to apply pressure to a fluid to cause its flow.

*Pumpak*　Standard pump control systems by Healy-Ruff Co.

**pump curves**　A set of graphical pump characteristics that represent pump performance by comparing total discharge head, net positive suction head, and efficiency relative to capacity.

**pump stage**　The number of impellers in a centrifugal pump.

**pump station**　A chamber that contains pumps, valves, and electrical equipment necessary to pump water or wastewater.

*Puratex*™　Cartridge filter by Osmonics, Inc.

*PurCycle*™　Volatile organic compound removal system by Purus Corp.

*Pureone*®　Laboratory water system by U.S. Filter Corp.

**pure oxygen process**　Variation of the activated sludge process using pure molecular oxygen for microbial respiration rather than atmospheric oxygen.

*Puresep*®　Resin separation process by U.S. Filter Corp.

*Purgamix*　Sludge mixing and heating system by Jones and Attwood, Inc.

*Purge Saver*™　Groundwater analyzer by QED Environmental Systems.

*Purifax*®　System using wet oxidation process for sludge stabilization by Leeds & Northrup.

**purified water**　See "USP-purified water."

*Puritan*　Fluid recovery treatment system for spent coolant and oils by SanTech, Inc.

*Purofine*™　Ion exchange resin by Purolite Co.

**PURPA**　Public Utilities Regulatory Policies Act.

**purple salt**　See "potassium permanganate."

*Pusher*™　Dewatering screw press by William R. Perrin, Inc.

**putrefaction**   The decomposition of organic matter by bacteria, fungi, and oxidation that results in the formation of noxious products and/or foul-smelling gases.

**putrescible**   Organic matter that is likely to result in a rotten, foul-smelling product as it undergoes decay or decomposition.

*Putzmeister*   Sludge cake pump marketed by Asdor.

**PV**   See "pervaporation."

**PVC**   See "polyvinyl chloride."

*PWMP®*   Pure water management program for outsourced water services by U.S. Filter Corp.

**pyrite**   A mineral containing sulfur and iron, frequently found in coal.

**pyrogen**   Cell material from bacteria that produce fevers in mammals.

**pyrolysis**   The chemical decomposition of a material by heating in oxygen-deficient conditions.

*Pyrospout*   Fluid bed combuster/incinerator by Process Combustion Corporation.

*Python Press*   High-pressure filtration and dewatering device by Waste Tech, Inc.

# Q

**QA**   Quality assurance.

**QC**   Quality control.

*QL-1™*   Ultraviolet disinfection system by Infilco Degremont, Inc.

*QLS*   Quick-lock sprocket by Budd Co.

*Q-Tracker™*   Sewer collection flow monitor by Badger Meter, Inc.

*Quadra-Clean™*   Vertical basket batch-type centrifuge by Western States Machine Company.

*Quadra-Kleen*   Sand filter backwash system by Culligan International Corp.

*Quadramatic III®*   Automatic batch-type centrifuge by Western States Machine Company.

*Quadra Press*   Plate and frame filter press by Duriron Co., Filtration Systems Division.

*Quadricell®*   Mechanical gas induction flotation separator by U.S. Filter Corp.

*Quadrufil*   Gravity filter system by Vulcan Industries, Inc.

**quagga mussel**   Freshwater mollusk that can foul water intake screens and piping by attaching itself to a solid structure, eventually restricting flow.

**qualitative**   The general description of a substance without specifying its exact amount or concentration.

*Quanti-Cult™*   Quality control test organisms by IDEXX Laboratories, Inc.

**quantitative**   Description of a substance in exact terms.

*Quanti-Tray™*   Water sample test kit by IDEXX Laboratories, Inc.

*Quantum*®   Floating aerator by Air-O-Lator Corp.

**quench**   To cool suddenly by immersion in water or oil.

**quicklime**   A calcium oxide material produced by calcining limestone to liberate carbon dioxide.

*Quick-Purge*®   Soil and groundwater remediation technology by Integrated Environmental Solutions, Inc.

*Quik-Clamp*   Clamp-on spray nozzle by FMC Corp., MHS Division.

# R

**rabble arms**   Rotating rake arms used to scrape sludge in a multiple-hearth furnace.

**race track**   See "oxidation ditch."

**RACT**   Reasonably available control technology.

**rad**   Radiation absorbed dose. A measure of the absorbed dose of radiation.

*Radial Filter*   Tertiary filtration system by Aero-Mod, Inc.

**radial flow**   Direction of flow, either from the center to the periphery or from the periphery to the center.

*Radial Plate Dryer*   Sludge drying system by Envirotech Systems Corp.

**radial well**   A well system where one or more cylindrical screens are driven horizontally into a water-bearing stratum, radiating from a central sump.

**radiation**   The transfer of energy by means of electromagnetic waves or high-speed particles.

**radiation sickness**   A sickness that results from overexposure to radiation, symptoms of which may include nausea, bleeding, hair loss, and death.

**radical**   A combination of atoms in a molecule that remains unchanged throughout most chemical reactions.

*Radicator*   Solid waste incinerator by Hitachi Metals America, Ltd.

**radioactive**   A description of an unstable atom that undergoes spontaneous disintegration and gives off radiant energy in the form of particles or rays.

**radioactivity**   The spontaneous decay or disintegration of an atomic nucleus that is accompanied by radiation.

**radionuclide**   An atom that spontaneously undergoes radioactive decay.

**radon**   A radioactive gas produced from the decay of radium that may be inhaled when the gas is released from groundwater, often during showering, bathing, or cooking.

**radon daughters**   Radioactive compounds produced during the decay of radon.

**radwaste**   Radioactive waste.

**raffinate**   The extracted waste stream containing contaminants in a solvent extraction process.

**railway softening**   See "excess lime-soda softening."

**rainforest**   See "tropical rain forest."

*Rainlogger*   Stormwater sampling unit by American Sigma.

*Rake-O-Matic*   Hydraulically operated, reciprocating rake bar screen formerly offered by BIF Division of General Signal.

*Ram*®   Waste compactor by S&G Enterprises, Inc.

*Ram-Rod*   Dewatering screw press by Katema, Inc.

*Ranney*   Well screen and caisson intake products by Hydro Group, Inc., Ranney Division.

*Ranney Intake*   Surface water intake system utilizing a passive screen/caisson arrangement by Hydro Group, Inc., Ranney Division.

**RAP**   Remedial action plan.

*RaPID Assay*™   Reagent kit for field soil analysis by Ohmicron.

*Rapid Decanter*®   Solid bowl scroll centrifuge by Flottweg.

*Rapid Gravity Dewatering*™   Inclined gravity filter for sludge dewatering by Wil-Flow, Inc.

**rapid mix** A physical water treatment process that involves rapid and complete mixing of coagulants or conditioning chemicals.

*Rapidor* Pressure leaf filter by Liquid-Solids Separation Corp.

**rapid sand filter** Granular media filter in which water flows downward through a sand filter bed at rates typically ranging from 80 to 320 L/min/m² (2 to 8 gpm/sq ft) of surface area.

**RAS** See "return activated sludge."

**rasp** A machine that grinds waste into a manageable size and controls odors.

**RATA** See "relative accuracy test audit."

*RatedAeration®* Circular steel activated sludge wastewater treatment plant by FMC Corp., MHS Division.

**rate-of-flow controller** A device that automatically controls the rate of flow of a fluid.

*Ratio* Turbidimeter by Hach Co.

*RatioFlo™* Flow-to-polymer ratio valve by Stranco.

**raw sludge** Undigested sludge recently removed from a sedimentation basin.

**raw water** Untreated surface water or groundwater.

*Raymond* Sludge incinerator formerly offered by Dorr-Oliver, Inc.

*Raypro* Air diffuser by Ray Products, Inc.

*Raysorb* Activated carbon volatile organic compound control system by RaySolv, Inc.

**RBC** See "rotating biological contactor."

*RCC* Resource Conservation Company.

**RCRA** See "Resource Conservation and Recovery Act."

**R&D** Research and development.

**RDF** See "refuse-derived fuel."

*Reacher* Reciprocating rake bar screen by Schloss Engineered Equipment.

**reactant** Any substance taking part in a chemical reaction.

**reaction rate** The rate at which a chemical reaction occurs.

**reactivation** The process of removing adsorbed organics and restoring the adsorptive characteristics of an adsorbent, usually by thermal or chemical means.

*Reactivator®* Solids contact clarifier by Graver Co.

**reactive wall** A permeable vertical wall constructed of a reactive mixture and installed below grade to treat groundwater that flows through it.

**reactive waste** A solid waste that is normally unstable and readily undergoes violent change, generates toxic gases or fumes, or is capable of detonation or explosion.

**reactor** The container or tank in which a chemical or biological reaction is carried out.

*Reactor-Thickener* Sludge thickening device using mixers and dewatering screens by Ralph B. Carter Co.

*React-pH™* Activated carbon by Calgon Carbon Corp.

**reagent** A chemical added to a system to bring about a chemical reaction.

**reagent grade water** High-purity water suitable for use in making reagents of for use in analytical procedures.

**reboiler** An evaporator-condenser unit that produces secondary steam after condensation of primary steam. Used to isolate steam systems and avoid cross-contamination of boiler chemicals.

**recalcining** Recovery of lime from water or wastewater sludge, usually with a multiple-hearth furnace.

**recarbonation** The reintroduction of carbon dioxide into water, usually during or after lime-soda softening.

**receiving water** Surface water body that receives effluent from a wastewater treatment plant.

**recharge** The natural or artificial process of replenishing an aquifer.

**reciprocating rake bar screen**   An automatic bar screen with a single rake that is raised and lowered to clean a stationary bar rack.

**reclaimed wastewater**   Wastewater treated to a level that allows its reuse for a beneficial purpose.

*Recla-Mate*®   Modular physical-chemical treatment plant by Wheelabrator Engineered Systems, Inc., Microfloc Products.

**reclamation**   The process of improving or restoring the condition of land or other material to a better or more useful state.

*Recla-Pac*   Package biological treatment plant by Wheelabrator Engineered Systems, Inc., Microfloc Products.

**recreational waters**   Any water body used for recreational activities such as swimming, boating, or fishing.

*Rectangulaire*   Package wastewater treatment unit consisting of a final clarifier and aeration unit by FMC Corp., MHS Division.

**recycle ratio**   The recycled flow rate divided by the influent flow rate in an activated sludge wastewater treatment system, or other process system.

**recycling**   The process by which recovered materials are transformed into new products.

**red bag waste**   Medical or infectious wastes.

*Red-B-Gone*®   Rust and iron stain remover by Technical Products Corp.

*Red Fox*   Sewage treatment systems for marine applications by Red Fox Environmental, Inc.

**Red List**   A list of 23 dangerous substances, designated by the U.K., whose discharge to the water should be minimized.

**redox potential**   See "oxidation-reduction potential."

*Red Rubber*™   Bar screen toothed rake segments of cast urethane by Rubber Millers, Inc.

**red tide**   A reddish discoloration of water caused by an excessive growth of certain microbes, whose toxins can cause massive fish kills.

**reducing agent**   Any substance that can give up electrons in a reaction.

**reduction**   A chemical reaction in which an element or compound gains electrons.

**red water**   Water whose reddish color usually results from the precipitation of iron salts or the presence of microbes that depend on iron or manganese.

*REECO*   Regenerative Environmental Equipment Co.

**reed**   Any of a variety of tall slender grasses grown in wet areas.

**reed bed**   Tertiary wastewater treatment system where organics remaining in secondary effluent are used as nutrients in the growth of reeds.

*REEF®*   Fine-pore floor-mounted diffuser by Environmental Dynamics, Inc.

*ReelAer*   Horizontal cage surface aerator formerly offered by Walker Process Equipment Co.

**reference dose (RfD)**   The exposure level of a carcinogenic contaminant thought to be without significant risk to humans when ingested daily over a specified time period.

*Refotex*   Fine-bubble diffuser by Refractron Technologies Corp.

*Refractite*   Ceramic filter membrane by Refractron Technologies Corp.

**refractory**   A highly heat resistant material used as a liner in a furnace or incinerator.

**refractory organics**   Organic substances that are difficult or impossible to metabolize in a biological system.

**refrigerant**   A substance that, by undergoing a phase change, lowers the temperature of its environment.

Commercial refrigerants, which include chlorofluoro-carbons and hydrofluorocarbons, are liquids whose latent heat of vaporization results in cooling.

**refuse**    All solid waste material discarded as useless.

**refuse-derived fuel (RDF)**    Fuel produced from municipal solid waste through shredding, pyrolysis, or other methods.

*Regal*™    Gas chlorinator by Chlorinators, Inc.

**regenerant**    A chemical used to restore the exchange capacity of ion exchange resin.

**regenerate**    The process of restoring exchange capacity of an ion exchange material.

**regenerative thermal oxidizer (RTO)**    An emissions control device that utilizes heat to accomplish volatile organic compound oxidation.

*Re-Gensorb*™    Volatile organic compound and hazardous air pollutant removal system by M&W Industries, Inc.

**Regional Administrator**    The administrative head of each of the ten regions organized by the U.S. EPA.

*Reg-U-Flo*®    Vortex valve flow control device by H.I.L. Technology, Inc.

*Rehydro-Floc*™    Aluminum chlorohydrate flocculant solution by Reheis, Inc.

**reject**    The waste stream containing impurities rejected in a treatment process, most commonly applied to reverse osmosis, electrodialysis, and ultrafiltration systems.

**reject staging**    Reverse osmosis process configuration where the reject from one stage is used as feedwater on a subsequent stage to increase water recovery. Also called "brine staging."

*REKO*    Bar screen products by Kopcke Industrie B.V.

**relative accuracy test audit (RATA)**    A comparison between an emissions stack tester's readings and readings obtained from a continuous emissions monitoring system.

**relative humidity**   The amount of water vapor present in the air, expressed as a percentage of the maximum amount that the air could hold at a given temperature.

**release**   Any occurrence where a regulated substance is discharged or emitted into the air, soil, or water.

**rem**   Roentgen equivalent man. A measure of the effective radiation dose absorbed by human tissue.

**remedial investigation and feasibility study (RI/FS)**   An evaluation of the risks associated with a hazardous waste site that includes the process of selection of an appropriate remedy.

**remediation**   The treatment of waste to make it less toxic and/or less mobile, or to contain a site to minimize further release.

**rendering plant**   A plant that converts grease and livestock carcasses into fats, oils, and other products.

**renewable resource**   A resource that theoretically cannot be totally consumed due to its ability to reproduce or regenerate.

*Renneburg*   Sludge dryer manufacturing division of Heyl & Patterson, Inc.

*Reo-Pure*   Reverse osmosis system by Great Lakes International, Inc.

*Reporter*™   Multiprobe for measuring water conditions by Hydrolab Corp.

**reservoir**   An artificial or natural pond, lake, basin, or tank that is used to store or control water.

**residence time**   The period of time that a volume of liquid remains in a tank or system.

**residue**   Solid or semisolid material remaining after processing, evaporation, or incineration.

**resin** A substance having ion exchange properties used in ion exchange demineralizers.

**resin beads** Spherical beads with ion exchange properties used in ion exchange demineralization.

**resistivity** A measure of resistance to the flow of electricity, used as an accurate measure of a water's ionic purity.

*Resi-Tech* Division of Aero-Mod, Inc.

**Resource Conservation and Recovery Act (RCRA)** 1976 U.S. law, amended in 1984, to regulate management and disposal of solid and hazardous wastes.

**respiration** Intake of oxygen and discharge of carbon dioxide as a result of biological oxidation.

**respirometer** An instrument used to study the character and extent of respiration.

*Retec®* Heavy metal recovery systems offered by Wheelabrator Engineered Systems, Inc., HPD Division.

**retention pond** A basin used for wastewater treatment and/or storage.

**retention time** The length of time that water or wastewater will be retained in a unit treatment process or facility.

*Re-Therm™* Thermal volatile organic compound oxidation unit by Regenerative Environmental Equipment Co.

*Retox* Regenerative volatile organic compound oxidizer by Adwest Technologies, Inc.

*Retroliner* Forms for filter underdrain rehabilitation by Roberts Filter Manufacturing Co.

**return activated sludge (RAS)** Settled activated sludge that is returned to mix with raw or primary settled wastewater.

**return sludge** See "return activated sludge."

**reverse osmosis (RO)** A method of separating water from dissolved salts by passing feedwater through a semipermeable membrane at a pressure greater than the osmotic pressure caused by the dissolved salts.

*Revolver*™   Rotary adsorption unit for volatile organic compound adsorption by Vara International.

*Rex*   Environmental equipment product line marketed by Envirex, Inc.

*Rex Chainbelt*   Former name of former parent company of Envirex, Inc.

**Reynolds number**   A nondimensional number that measures the state of turbulence in a fluid system. It is calculated as the ratio of inertia effects to viscous effects.

**RF/AS**   Roughing filter/activated sludge.

**RfD**   See "reference dose."

**RFP**   Request for proposal.

**RFQ**   Request for quotation.

**RHA**   See "Rivers and Harbors Act."

**rheology**   The study of the deformation and flow of substances.

**rhizosphere**   The zone of intermingled roots and soil.

**ribonucleic acid (RNA)**   One of two types of long-chain molecules containing hereditary material vital to reproduction.

*Richards of Rockford*   Former equipment manufacturer acquired by Aqua-Aerobic Systems, Inc.

*Rich Tech*   Equipment product line by Aerators, Inc.

**RI/FS**   See "remedial investigation and feasibility study."

*RIGA*   Former equipment manufacturer.

*Riga-Sorb*™   Activated carbon adsorbers by Farr Co.

*Rim-Flo*   Peripheral feed, circular clarifier by Envirex, Inc.

**Ringleman test**   A method of quantifying the opacity of a stack plume by comparing it to a set of standard disks having increasing degrees of discoloration.

*RingSparjer*   Air injection diffuser by Walker Process Equipment Co.

**riparian rights**   A landowner's rights to the water on or bordering his or her property, including the right to prevent diversion or misuse of it upstream.

**rip rap**   Stone or rocks placed to protect dams, levees, or dikes to protect from wave action.

**rise rate**   See "overflow rate."

**rising film evaporator**   An evaporator using vertical heat transfer surfaces where liquor on one side of the surface is boiled by steam condensing on the other side, causing vapors to rise, carrying the liquid upward as a film.

*RKL®*   Pinch valve by Robbins & Myers, Inc.

**R-MAP**   Regional Management Assessment Program.

*RMP*   Roberts Manhattan Process.

**RNA**   See "ribonucleic acid."

**RO**   See "reverse osmosis."

*Robo™*   Bar screen by Vulcan Industries, Inc.

*Robo Rover™*   Traversing bar screen by Vulcan Industries, Inc.

*Robo Stat™*   Stationary bar screen by Vulcan Industries, Inc.

**rodding**   A method of cleaning tubes or sewers using long rods that are able to remove or dislodge debris.

*Roebelt*   Belt filter press by Roediger Pittsburgh, Inc.

*Roedos*   Mixing system for dry and liquid polymers by Roediger Pittsburgh, Inc.

*Roefilt*   Sieve drum concentrator by Roediger Pittsburgh, Inc.

*Roeflex*   Fine-bubble diaphragm diffuser by Roediger Pittsburgh, Inc.

*Roemix*   Lime post treatment mixing system for dewatered sludge by Roediger Pittsburgh, Inc.

*Roepress*   Belt filter press by Roediger Pittsburgh, Inc.

*Ro-Flo®*    Sliding vane compressor by A-C Compressor Corp.

**ROG**    Reactive organic gases.

*Rogun*    Reverse osmosis membrane cleaning solution by Argo Scientific.

*RollAer*    Aerobic digestion aeration equipment by Walker Process Equipment Co.

*Roll-Dry*    Internally fed rotary fine screen by Schlueter Co.

*Rolling Grit*    Aerated tank grit washing and removal unit by Walker Process Equipment Co.

*Romembra*    Reverse osmosis membrane elements by Toray Industries, Inc., and Ropur AG (Europe).

*Romicon®*    Hollow fiber membrane filtration products by Koch Membrane Systems, Inc.

*Romi-Kon™*    Oil-water emulsion separator by Koch Membrane Systems, Inc.

*Roots*    Centrifugal compressor and blower product line by Dresser Industries/Roots Division.

*Roplex*    Live bottom feeder for solids and storage piles by Hindon Corp.

*Roptic®*    Filter cake sensor by Rosenmund.

*Rosep™*    Reverse osmosis systems by Graver Co.

*Rotadisc®*    Sludge dryer by Stord, Inc.

*Rotafine*    Rotary fine screen by Jones and Attwood, Inc.

*Rotamat®*    Screening equipment product line by Lakeside Equipment Corp. (U.S.) and Hans Huber (Europe).

**rotameter**    A variable area liquid flowmeter.

*Rotapak*    Screw-type screenings compactor by Longwood Engineering Co., Ltd.

*Rota-Rake*    Circular sludge collector by Graver Co.

*Rotarc*    Arc-type bar screen by John Meunier, Inc.

**rotary collector**    Rotating mechanisms used in circular clarifiers to collect and remove settled solids.

**rotary distributor**   Rotating pipe that evenly distributes wastewater on the surface of a trickling filter.

**rotary drum screen**   Cylindrical screen used to remove floatable and suspended solids.

**rotary drum thickener**   Rotating cylindrical screen used to thicken sludge.

**rotary kiln**   An incinerator consisting of a slowly rotating, long horizontal cylinder in which material is fed at one end and tumbled by the kiln to promote drying as it is conveyed to the other end.

*Rotasieve*   Externally fed rotary fine screen by Jones and Attwood, Inc.

**rotating biological contactor (RBC)**   A fixed-film biological treatment device where biological growths are grown on circular discs, mounted on a horizontal shaft that slowly rotates through wastewater.

*Rotex*   Rotating grit removal system by Simon-Hartley, Inc.

**rotifer**   A very small aerobic, multicellular animal that feeds on organic matter in wastewater.

*Rotobelt*   In-channel fine screen by Dontech, Inc.

*Roto-Brush*   Rotary screen brush cleaning device by Dontech, Inc.

*Roto-Channel*   Combination bar screen and compacting conveyor by Dontech, Inc.

*RotoClean*   Screenings washer by Parkson Corp.

*RotoClear*   Microscreen formerly offered by Walker Process Equipment Co.

*Rotoco®*   Continuous duty granular media filter by Eimco Process Equipment Co.

*Rotodip®*   Volumetric feeder for liquids or slurries by Leeds & Northrup.

*RotoDip*   Manually controlled slotted pipe skimmer by Walker Process Equipment Co.

*Rotodisintegrator*   Debris grinder by Zimpro Environmental, Inc.

*Roto-Drum*   Internally fed rotary fine screen and thickener by Dontech, Inc.

*Rotofilter*   Externally fed rotary fine screen by Sepra Tech.

*Roto-Guard®*   Horizontal drum screen/thickener by Parkson Corp.

*Rotoline*   Rotary distributor for trickling filter by FMC Corp., MHS Division.

*Rotomite*   Sludge handling dredger by Crisafulli Pump Co.

*Rotopac®*   Screw-type screenings compactor by John Meunier, Inc.

*Rotopass*   Externally fed rotary fine screen by Passavant-Werke AG.

*Rotopress®*   Shaftless screw compactor by Andritz-Ruthner, Inc.

*RotoPress*   Screenings compactor by Parkson Corp.

*Roto-Press*   Screenings compactor by Roto-Sieve AB.

*Roto-Press*   Combination rotary fine screen and dewatering press by Dontech, Inc.

**rotor**   See "brush aerator."

*Rotordisk*™   Rotating biological contactor by CMS Group, Inc.

*Rotorobic*   Package rotating biological contactor system by Hycor, Inc.

*Rotoscoop®*   Self-cleaning volumetric feeder by Wyssmont Co., Inc.

*Roto-Scour*   Sand filter underdrain system by Graver Co.

*RotoSeal*   Rotary distributor for trickling filters by Walker Process Equipment Co.

*Roto-Sep*™   Primary wastewater treatment system by Dontech, Inc.

*Rotoshear* Internally fed rotary fine screen by Hycor Corp.

*Roto-Sieve®* Internally fed rotary fine screen by Roto-Sieve AB.

*Roto-Skim* Rotary pipe skimmer by Envirex, Inc.

*Rotosludge®* Drum-type rotary sludge thickener by Hycor Corp.

*Rotospir®* Shaftless screw conveyor by Andritz-Ruthner, Inc.

*Rotostep* In-channel bar screen by Hycor Corp.

*Rotostrainer* Externally fed rotary fine screen by Hycor Corp.

*Rotosweep* Filter media surface agitator by Roberts Filter Manufacturing Co.

*Rototherm®* Agitated thin film evaporator by Artisan Industries, Inc.

*Roto-Thickener™* Rotary drum sludge thickener by Dontech, Inc.

*Roto-Trak* Sludge plows, or chicanes used with gravity sludge dewatering by Komline-Sanderson.

*Roto-Trols* Pressure-operated pump controller by Healy-Ruff Co.

*Rotox* Submersible aeration and mixing system by Davis Water & Waste Industries, Inc.

**roughing filter** A high-rate filter designed to receive high hydraulic or organic loading rates as a first, or intermediate, treatment step.

*Rover* Rust removing compounds by Hach Company.

**ROWPU** Reverse osmosis water purification unit.

*Roxidizer®* Volatile organic compound and air toxics controls installation by Tellkamp Systems, Inc.

*Royce Process Equipment* Former manufacturer of wastewater treatment equipment.

**rpm** Revolutions per minute.

**RQ**   Reportable quantity.

*RSDS*   Vacuum-assisted rapid sludge dewatering system by the former U.S. Environmental Products, Inc.

**RTO**   See "regenerative thermal oxidizer."

**RTR**   Reinforced thermosetting resin.

**rubbish**   Combustible and noncombustible solid waste from residential and commercial sources.

**run**   The time period or continuous course during which a unit operates or a test occurs.

**rundown screen**   See "static screen."

**runoff**   Rainwater, leachate, or other liquid that drains over land and reaches a drain, sewer, or body of water.

**rupture disk**   A diaphragm designed to burst at a predetermined pressure differential.

**Ryznar Index**   A scale used to evaluate the corrosion or scaling potential of water.

# S

*SAB Reactor*  Package wastewater treatment plant by Biosab, Inc.

**SAC**  Starved air combustion.

*S.A.C.*™  Sludge age control system by United Industries, Inc.

**sacrificial anode**  A sacrificial piece of metal, usually zinc or magnesium, electrically connected to a more noble metal in an electrolyte. The anode goes into solution at an accelerated rate to protect the more noble metal from corrosion.

**SAE**  Society of Automotive Engineers.

**Safe Drinking Water Act (SDWA)**  U.S. Act that ensures public water supplies are free of contaminants that may cause health risks and prevents endangerment of underground sources of drinking water.

*Safgard*  Rotary fine screen products by Schlueter Co.

**salinity**  The presence of soluble minerals in water.

**salinization**  The accumulation of salts in a soil to the extent that plant growth is inhibited, usually occurring as a result of excessive irrigation in an arid area.

**Salmonella**  Aerobic bacteria that is pathogenic in humans and chiefly associated with food poisoning.

**salmonellosis**  A common type of food poisoning characterized by a sudden onset of gastroenteritis caused by eating food contaminated with Salmonella bacteria.

**salt**  A class of ionic compounds formed by the combination of an acid and a base.

**salt flux**   The amount of dissolved substances that pass through a reverse osmosis membrane.

**salt marsh**   A coastal marsh that is periodically flooded with salt water.

*Saltmaster*   Water softener brine reclamation system by Bruner Corp.

**salt pan**   An accumulation or layer of salts in the soil that may be toxic to agricultural crops.

**salt rejection**   In reverse osmosis, the ratio of salts removed to the original salt concentration.

**salt splitting**   The conversion of salts to their corresponding acids.

**salt water**   Water containing a dissolved salt concentration ranging from 10,000 to 35,000 mg/L.

**salt water intrusion**   The intrusion of salt water into a body of fresh surface water or groundwater.

*SAM*™   Status alert modem to monitor disinfectant dosing system by Strantrol, Inc.

**sampling well**   See "monitoring well."

*Sand Dollar*   Sludge harvesting machine for sludge drying beds by Cherrington Corp.

**sand drying bed**   See "sludge drying bed."

**sand filter**   See "granular media filter."

*Sandfloat*   Combination dissolved-air flotation and sand filter treatment system by Krofta Engineering Corp.

*SandPIPER*®   Diaphragm pump by Warren Rupp, Inc.

*Sandsep*®   Screw-type grit classifier by Spirac AB (Europe) and JDV Equipment Corp. (U.S.).

*Sandwash*   Hydrocyclone by Serck Baker, Inc.

*Sanilec*   Sodium hypochlorite generating systems by Eltech International Corp.

*Sanilo*™   Water treatment product line by U.S. Filter Corp.

*Sani-Sieve*  Gravity-fed static screen by Dontech, Inc.

**sanitary connection**  A single family residential connection, or single commercial or industrial connection, to a public water supply system.

**sanitary landfill**  A landfill for disposing of solid wastes.

**sanitary sewer**  Collection system of underground piping used to remove sanitary wastewater.

**sanitary sewer overflow (SSO)**  Overloaded operating condition of a sanitary sewer that results from inflow/infiltration.

**sanitary wastewater**  Domestic wastewater, without storm and surface runoff, that originates from sanitary conveniences.

*San-I-Tech™*  Grease interceptor by Scienco/Fast Systems.

*Sanitron™*  Ultraviolet water purifier by Atlantic Ultraviolet Corp.

*Sanuril®*  Hypochlorite tablet disinfection system by Eltech International Corp.

**saponify**  The conversion of a fat or grease into a soap by reaction with an alkali.

**saprophytic bacteria**  Bacteria that feed on dead or nonliving organic matter.

**SARA**  See "Superfund Amendments and Reauthorization Act."

*Satellite*  Electrically driven rotary distributor for fixed-film reactor by Simon-Hartley, Ltd.

*Sation®*  Water treatment product line by U.S. Filter Corp.

**saturated steam**  Vapor in equilibrium with water at the boiling temperature, containing no liquid.

**saturated zone**  See "zone of saturation."

**saturation**  The maximum concentration of a phase or material that can be contained within another phase or material.

**Saturation Index**   See "Langelier Saturation Index."

*SAV 715*   Stainless steel sludge collector chain by Hitachi Maxco, Ltd.

**save-all**   Separation device used in a paper mill to reclaim fibers and fillers from white water.

*Save-All*   Clarifier designed for paper mill fiber recovery by Walker Process Equipment Co.

**SBA**   Strong-base anion exchanger.

**SBC**   Submerged biological contactor.

**SBOD**   Soluble BOD. See "biochemical oxygen demand."

**SBR**   See "sequencing batch reactor."

**SBS**   See "sick building syndrome."

*SC™*   Package spray-type deaerating heater by Graver Co.

**SCADA**   Supervisory control and data acquisition.

*SCADA-Flo™*   Open-channel transmitter by Marsh-McBirney.

**scale**   A precipitate that forms on a surface as a result of a physical or chemical change.

**scanning electron microscope (SEM)**   A microscope with a magnification range from 20 to 200,000× at a resolution of 100 Å, where illumination is provided by a beam of electrons that scan the specimen surface.

**scarp**   A steep, almost perpendicular slope.

*SCAT™*   Secondarily contained aboveground tank by Industrial Environmental Supply, Inc.

*Scavenger*   Robotic scrubber and vacuum sludge removal system by Aqua Products, Inc.

**scavenging**   (1) The unauthorized and/or uncontrolled removal of materials at any point in a solid waste management system. (2) The removal of a substance by converting it to another form or adsorbing it onto another compound.

**SCBA**   Self-contained breathing apparatus.

**SCD**   See "streaming current detector."

*Scentoscreen*    Portable gas chromatograph for volatile organic compound analysis by Sentex Systems, Inc.

**SCFM**    Standard cubic feet per minute.

**Schistosoma**    A flatworm or blood fluke that is highly parasitic to snails during one phase of its life and to humans during another.

**schistosomiasis**    A waterborne disease of tropical and subtropical regions transmitted to humans who wade or bathe in water infested by Schistosoma, with freshwater snails acting as intermediate hosts.

**schmutzdecke**    A biologically active layer that forms on the top of slow sand filters to aid in the removal of suspended solids.

**Schoop process**    Process for coating steel that uses a blast of air to spray a mist of molten metal onto the surface to be protected.

**scientific method**    An orderly method of obtaining, organizing, and applying new knowledge.

*Scion*®    Short-cycle ion exchange system by U.S. Filter Corp.

*SCONOx*™    Combined NOx/CO catalytic absorber system by Goal Line Environmental Technologies.

*Scoop-A-Fish*    Traveling water screen fish collection trough by Norair Engineering Corp.

*ScorGuard*®    Organic cooling tower water treatment additive by Western Water Management, Inc.

**scouring velocity**    The minimum velocity required to carry away material accumulations in a conduit or pipeline.

*Scour-Pak*®    Granular media gravity depth filter by Graver Co.

**SCR**    See "selective catalytic reduction."

**SCR**    Silicon controlled rectifier.

**screening**  (1) A treatment process using a device with uniform openings to retain coarse solids. (2) A preliminary test method used to separate according to common characteristics.

**screenings**  The material removed by a mechanical screening device.

**screenings press**  A mechanical press used to compact and/or dewater material removed from mechanical screening equipment.

*Screezer*  Combination screening and dewatering device by Jones and Attwood, Inc.

**screw conveyor**  A conveyor utilizing a helical screw rotating within a trough to convey material.

*Screwpeller*™  Centrifugal screw impeller used in surface aerator by Aeration Industries, Inc.

**screw pump**  A low-lift, high-capacity pump that raises water by means of a slowly rotating inclined shaft fitted with a helical blade that revolves in a trough or pipe.

**scrubber**  A device used to remove particulates or pollutant gases from combustion or chemical process exhaust streams.

**scrubbing**  The removal of impurities from an air or gas stream by entraining the pollutants in a water spray.

*ScruPac*™  Screw-type screenings compactor by Vulcan Industries, Inc.

*Scru-Peller*®  Sludge pump by Yeomans Chicago Corp.

*SCUBA*™  Self-contained gate and valve actuator by Rodney Hunt Co.

**scum**  Floatable materials found on the surface of primary and secondary settling tanks consisting of food wastes, grease, fats, paper, foam, and similar matter.

**scum breaker**  A device installed in a sludge digester to break up scum.

*Scumbuster*™   Pump used to chop solids in a digester scum blanket by Vaughn Company, Inc.

**scum collector**   A mechanical device for removing scum from the surface of a settling tank.

*Scum Sucker*™   Telescopic pipe for scum removal by United Industries, Inc.

**scum trough**   A trough used in a primary sedimentation basin to remove scum and convey it from the basin.

**SCWO**   See "supercritical water oxidation."

**SDI**   See "silt density index."

**SDWA**   See "Safe Drinking Water Act."

*Sea Devil*   Floating oil skimmer by Vikoma International.

**sea lettuce**   Common seaweed that can grow in nuisance concentrations in the presence of excess nutrients.

**sea level**   The average surface level of the sea, uninfluenced by tidal movement or waves; used as a reference for elevation.

*Sealtrode*   Sealed pump controller by Yeomans Chicago Corp.

*Seaskimmer*   Floating oil skimmer by Vikoma International.

**seawall**   A wall built to protect a coastline from erosion caused by wave action.

**seawater**   General term for sea or ocean water, with a typical total dissolved solids concentration of 35,000 mg/L.

**Secchi disk**   A small disk, divided into black and white quadrants, that is lowered into water to visually observe water clarity and estimate the depth of the euphotic zone.

**Secchi disk depth**   The water depth at which a Secchi disk is no longer visible. In a lake, this depth is approximately equal to the euphotic zone.

**secondary clarifier**   A clarifier following a secondary treatment process, designed for gravity removal of suspended matter.

**secondary contaminant**   A contaminant that affects drinking water taste, odor, or aesthetics.

**secondary effluent**   Treated wastewater leaving a secondary treatment facility, usually having a "$BOD_5$" and suspended solids of less than 30 mg/L.

**secondary emissions**   Emissions that occur as a result of the construction or operation of a facility but do not come from the facility.

**secondary MCL**   EPA-mandated maximum contaminant level in drinking water based on taste, odor, or aesthetics.

**secondary pollutant**   A pollutant formed in the environment as the result of a chemical reaction of two or more primary pollutants or naturally occurring elements.

**secondary sludge**   The sludge from the secondary clarifier in a wastewater treatment plant.

**secondary treatment**   The treatment of wastewater through biological oxidation after primary treatment.

**second-order reaction**   A reaction in which the rate of change is proportional to the square of the concentration of one of the reactants or to the product of the concentrations of two different reactants.

**second-stage biochemical oxygen demand**   See "nitrogenous oxygen demand."

**second-stage BOD**   See "nitrogenous biochemical oxygen demand."

**secure landfill**   Landfill that segregates and isolates hazardous materials from contacting each other, the groundwater, or the atmosphere.

*Sedifloat*   Water and wastewater treatment unit by Krofta Engineering Corp.

*Sediflotor®*   Dissolved-air flotation unit by Infilco Degremont, Inc.

**sediment**   The solid material that settles from a liquid.

**sedimentation**   The removal of settleable suspended solids from water or wastewater by gravity in a quiescent basin or clarifier.

**sedimentation basin**   A quiescent tank used to remove suspended solids by gravity settling. Also called clarifiers or settling tanks, they are usually equipped with a motor-driven rake mechanism to collect settled sludge and move it to a central discharge point.

*SEE*   Schloss Engineered Equipment, Inc.

**seed**   (1) Crystalline particles added to a supersaturated solution to induce precipitation. (2) Well-digested sludge used to seed a sludge digester.

**seepage pit**   A covered excavation that receives septic tank effluent and permits its effluent to seep through the bottom and sides of the excavation.

**seepage spring**   A spring occurring where the water table breaks the ground surface. Also called a "gravity spring."

*Seghers Pelletech®*   Indirect sludge dryer and pelletizing unit by Wheelabrator Clean Water Systems, Inc., Bio Gro Division.

**selective catalytic reduction (SCR)**   Flue gas treatment process for the removal of nitrogen oxides by reduction with ammonia to form elemental nitrogen and water.

*Selectofilter*   Revolving-drum screen strainer by FMC Corp., MHS Division.

*Selectostrainer*   In-line strainer by FMC Corp., MHS Division.

*Selemion*   Ion exchange membranes by Ashai Glass America, Inc.

*Selex®*   Graded-density cartridge filters by Osmonics, Inc.

*SelRO®*   Membrane filtration systems by LCI Corp.

**SEM**   See "scanning electron microscope."

**semipermeable**   A membrane that does not have measurable pores, but through which smaller molecules can pass.

**senescent lake**   A very old lake, almost full of sediment and rooted water plants, that will eventually become a marsh.

**sensible heat**   Heat measurable by temperature alone.

**sensitivity**   The ability of a unit or instrument to respond to a small difference in values.

*Sentinel*   Filter backwash control system by Roberts Filter Manufacturing Co.

*Sentre-Fier*   Rotary fine screen by Dontech, Inc.

**Sentry**   Groundwater gasoline recovery system by Douglas Engineering.

*SEPA®*   Reverse osmosis membranes by Osmonics, Inc.

**sepralators**   Membrane elements.

**septage**   Sludge produced in individual on-site wastewater treatment systems including septic tanks and cesspools.

**septic**   Condition characterized by bacterial decomposition under anaerobic conditions.

**septicity**   The condition that results from biological degradation of organic matter in wastewater under anaerobic conditions, usually producing hydrogen sulfide or other odorous compounds.

**septic tank**   A wastewater treatment system principally used for individual residences that combines sedimentation, sludge digestion, and sludge storage in a single or dual compartmented tank.

**septum**   A permeable material used to support filter media.

**sequencing batch reactor (SBR)** Treatment process characterized by the interruption of flow to the reactor during the sedimentation and decanting phase of treatment.

*Sequest-All* Sequestering agent used to control iron scaling and corrosion by Sper Chemical Corp.

**sequestering agent** An organic compound, such as ethylenediaminetetraacetic acid, that chemically sequesters other compounds or ions so they cannot be involved in chemical reactions.

**sequestration** The formation of a stable, water-soluble complex with an ion in solution to prevent precipitation or scaling.

*Seral®* Laboratory water treatment product line by U.S. Filter Corp.

*Serfilco®* Wastewater treatment equipment products by Serfilco, Inc.

**serogroup** A group of closely related organisms having one or more common antigens.

*Serpentix®* Convoluted, self-cleaning belt conveyor by Serpentix Conveyor Corp.

**service factor (SF)** A multiplier that, when applied to the rated power, indicates the permissible power loading that may be carried under the conditions specified.

*Sessil®* Polyethylene strip media for trickling filters by NSW Corp.

*SETLdek* Clarifier tube settlers by Munters.

**setpoint** An input value to be maintained by a control device.

**settleability** The tendency of suspended solids to settle.

**settleable solids** That portion of suspended solids that are of a sufficient size and weight to settle to the bottom of an Imhoff cone in 1 hr.

**settled sludge volume (SSV)**  Volume of settled sludge measured at predetermined time increments for use in process control calculations.

**settling tank**  A quiescent tank used to remove suspended solids by gravity settling. Also called "clarifiers" or "sedimentation basins," they are usually equipped with a motor-driven rake mechanism to collect settled sludge and move it to a central discharge point.

**settling tubes**  See "tube settlers."

**settling velocity**  The rate at which a particle settles through air or water.

**sewage**  See "wastewater."

**sewer**  Collection system of underground piping used to remove wastewater.

**sewerage**  The entire system of wastewater collection, treatment, and disposal.

*Sewer Chewer*™  Comminuter/sludge grinder by Yeomans Chicago Corp.

**sewer gas**  A gas mixture produced by anaerobic decomposition of organic matter usually containing high percentages of methane and hydrogen sulfide.

**sewershed**  Land area that drains into a sewer.

*Sewpadisc*  Rotating biological contactor by Biwater Treatment Ltd.

**SF**  See "service factor."

*SFT*™  Sediment flushing tank by John Meunier, Inc.

*ShallowTray*  Aeration system for removal of volatile organics by North East Environmental Products, Inc.

*Shann-No-Corr*  Zinc metaphosphate corrosion inhibitor and sequestering agent by Shannon Chemical Corp.

*Sharples*®  Division of Alfa Laval Separation, Inc.

*Sharpshooter*  Polymer feed and control system by Norchem Industries.

*Shearfuser*   Cast iron diffuser for anaerobic digestion by FMC Corp., MHS Division.

**sheet flow**   Overland stormwater flow in a thin sheet of uniform thickness.

**shell-and-tube heat exchanger**   A tubular heat exchanger housed within the shell of a pressure vessel.

**sherardizing**   A process for protecting iron from corrosion by means of a corrosion-resistant layer of zinc on the iron surface.

**Shigella**   A bacterium associated with dysentery that is transmitted through consumption of water or food contaminated with fecal matter.

**shock load**   A sudden increase in hydraulic or organic loading to a treatment plant.

**shore**   (1) The land bordering a body of water. (2) To brace or give support.

**short circuiting**   Uneven flow through a vessel that results from density currents or inadequate mixing that allows some currents to leave the vessel more quickly than others.

**short-term exposure limit (STEL)**   The   maximal allowable level of a material in workplace air, usually measured over a 15-min period.

*Shriver*®   Plate and frame filter press by Eimco Process Equipment Co.

**shute**   The horizontal wire in woven wire mesh, also called the "weft" wire.

**SIC**   See "Standard Industrial Classification."

**sick building syndrome (SBS)**   Condition in some buildings where occupants experience an undue number of symptoms of eye, skin, or nasal irritations, or lethargy, headache, and odor/taste complaints.

**side hill screen**   See "static screen."

**sidewall**  The wall at the side of a structure.

*Sidewall Separator*  In-channel clarifier for oxidation ditch by Lakeside Equipment Co.

**side water depth**  The depth of water measured along a vertical interior wall of a basin or tank.

*Side Winder*  Environmental screens by Cook Screen Technologies, Inc.

**sieve analysis**  A size distribution analysis filter sand sample using a series of standard sieve screens.

**sieve size**  The standard sieve size through which a sample of sand will pass.

*SightWell*  Circular clarifier with hydraulic suction type sludge removal system by Walker Process Equipment Co.

*Sigma*  Low-speed surface aerator by Purestream, Inc.

*Sigma Flight*  Fiberglass sludge collector flight by Envirex, Inc.

*SIHI-Halberg*  Digester draft tube sludge mixer by SIHI Pumps, Inc.

**silica**  A mineral composed of silicon and oxygen.

**silicate**  Any compound containing silicon, oxygen, and one or more metallic compounds.

**siliceous**  Composed of or containing silica or a silicate compound.

**silicosis**  A lung disease caused by prolonged inhalation of silica dust, which results in fibrosis or scarring of lung tissue.

*Sil-Kleer®*  Filter aid by Silbrico Corp.

**silo**  A tall, cylindrical storage vessel for dry solids.

*Silo Pac*  Chemical feed system by Wallace & Tiernan, Inc.

**silt**  Individual mineral particles ranging in size between fine sand and clay.

**silt density index (SDI)**  A measure of the fouling tendency of water based on the timed flow of a liquid through a membrane filter at a constant pressure.

**silting**  The deposition of silt or sediment in a water body.

*Silverback*™  Aqueous cleaner recovery system using ceramic membrane filters by U.S. Filter Corp.

*Silver Band*  Granular media pressure filter by Envirotech Company.

*Silverseries*™  Packaged desalination system by Matrix Desalination, Inc.

*Simcar*®  Turbine aerator by United Industries.

*Simon Hydro-Aerobics*  Former name of Hydro-Aerobics, Inc.

*Simon Waste Solutions*  Former name of Waste Solutions.

*Simon WTS*  Former name of Wastewater Treatment Systems, Inc.

*Simplex*  Low-speed surface mechanical aerators by Asdor Division of Edward & Jones, Inc.

*Simrake*  Rotary bar screen by Simon-Hartley, Ltd.

*Simspray*  Rectangular distributor for fixed-film reactor by Simon-Hartley, Ltd.

*Simultech*™  Biological nutrient removal system by Schreiber Corp.

*Sinclair*  Internally fed rotary fine screen by Bielomatik London, Ltd.

**SIP**  See "State Implementation Plan."

**SITE**  See "Superfund Innovative Technology Evaluation."

*Sitepro*™  Remediation site control system by ORS Environmental Equipment.

**SI unit**  The international system of units (Système International), largely based on the metric system, for measuring length, mass, volume, and radiation.

**skid mounted**  Equipment or equipment packages mounted on a horizontal structure or platform to facilitate handling and/or installation.

*Skim-Kleen®*  Oil removal belt skimmer by Tenco Hydro, Inc.

*Skim-Pak™*  Floating, self-adjusting weir skimmer by Douglas Engineering.

**skip**  Bar screen cleaning rake.

*SKRAM*  Acoustic fish behavioral control device by FMC Corp., MHS Division.

*S&L*  Smith & Loveless, Inc.

**slag**  Waste residues produced by metal smelting and coal gasification.

**slake**  The process of mixing lime with water to accomplish a chemical combination.

**slaked lime**  See "hydrated lime."

**slash and burn**  An agricultural practice where vegetation is cut, allowed to dry, and burned prior to soil cultivation and planting.

*Slide Gate*  Screenings press by Andritz-Ruthner, Inc. (Western Hemisphere) and Contra-Shear Engineering, Ltd.

**slimicide**  A substance used to prevent, inhibit, or destroy biological slimes.

*Slo-Mixer*  Axial flow paddle flocculators by Envirex, Inc.

**slop oil**  Separator skimmings and tramp oil generated during refinery startup, shutdown, or abnormal operation.

**sloughing**  The detachment of accumulated biological solids from trickling filter media.

**slow sand filter**  Sand filter characterized by low flow rates; it relies on the formation of a layer of solids on the top of the sand bed to accomplish most of the filtration.

**sludge**    Accumulated and concentrated solids removed from water and wastewater treatment processes, especially sedimentation basins.

**sludge age**    The average time that a microbial cell remains in an activated sludge system. It is equal to the mass of cells divided by the rate of cell wasting from the system.

*Sludge Age Controller*™    Sludge age control system by United Industries, Inc.

**sludge blanket**    The accumulated sludge hydrodynamically suspended in a solids-contact unit.

**sludge bulking**    A phenomenon that occurs in activated sludge plants where a sludge does not readily settle or concentrate.

*Sludgebuster*™    Sewage shredder by International Shredder, Inc.

**sludge cake**    Dewatered residue from a filter press, centrifuge, or other sludge dewatering device.

*SludgeCleaner*®    Sludge and scum screen and compactor by Parkson Corp.

**sludge collector**    Mechanisms used in clarifiers to collect and remove settled solids from the tank bottom.

**sludge conditioning**    See "conditioning."

*Sludge Detention Optimizer*™    Sludge thickening/conditioning process by Dontech, Inc.

**sludge dewatering**    The removal of a portion of the water contained in sludge by means of a filter press, centrifuge, or other mechanism.

**sludge digestion**    See "digestion."

**sludge dryer**    A device utilizing heat for the removal of a large portion of water within sludge.

**sludge drying bed**    A partitioned area consisting of sand or other porous material upon which sludge is dewatered by drainage and evaporation.

*Sludge Expert*   Automatic belt press control and management system by Solids Technology International Ltd.

*Sludge Gun*®   Sludge level detector by Markland Specialty Engineering, Ltd.

*Sludge Guzzler*   Hydraulically driven sludge pump by Guzzler Mfg., Inc.

*SludgeMaster*   Submersible air-powered pump by Warren-Rupp, Inc.

*Sludgepactor*   Sludge screen and compactor by Jones and Attwood, Inc.

*SludgePress*™   Belt filter press by Enviroquip, Inc.

**sludge stabilization**   Treatment process to convert sludge to a stable product for ultimate disposal or use, and reduce pathogens to produce a less odorous product.

**sludge volume index (SVI)**   The volume in milliliters occupied by 1 gram of settled sludge after settling for 30 min in a graduated cylinder.

*Sludgifier*   Lagoon dredge by VMI Inc.

*Sludglite*   Sludge blanket level detector by Ecolotech Corp.

**slug load**   A sudden hydraulic or organic load to a treatment plant.

**sluice gate**   Manual or power-operated gate used to isolate a channel from flow.

**slurry**   A suspension of a relatively insoluble chemical in water, usually having a suspended solids concentration of 5000 mg/L or more.

*Slurrystore*   Slurry storage system by A.O. Smith Harvestore Products, Inc.

**small quantity generator (SQG)**   A generator of between 100 and 1000 kg/month of hazardous wastes.

*SmartFilter*™   Traveling bridge filter by Agency Environmental, Inc.

**SmartRO™** Reverse osmosis system by Water and Power Technologies.

**SMCRA** See "Surface Mining Control and Reclamation Act."

**SM-Cyclo®** Speed reducer and gearmotor product line by Sumitomo Machinery Corp.

**smog** A type of air pollution characterized by reduced visibility due to atmospheric particulates and elevated levels of photochemical oxidants.

**Smogless™** Wastewater treatment, sludge drying, and incineration plants by U.S. Filter Corp.

**smoke** The suspended matter in an exhaust emission that obscures the transmission of light.

**smoke number** A dimensionless term quantifying smoke emissions.

**Smooth-tex** Rectangular woven mesh for screening equipment by Envirex, Inc.

**SMX** Belt filter press by Andritz-Ruthner, Inc.

**SN** See "smoke number."

**Snail** Grit dewatering system by Eutek Systems.

**Snowflake Packing** Plastic packing media for air stripping applications by Norton Co.

**SO₂** Sulfur dioxide.

**SOC** Synthetic organic chemicals.

**SOCMI** Synthetic organic chemical manufacturing industry.

**soda ash** See "sodium carbonate."

**sodium aluminate** An auxiliary coagulant used in water treatment. Chemical formula is $Na_2Al_2O_4$.

**sodium bisulfite** A liquid dechlorinating agent. Chemical formula is $NaHSO_3$.

**sodium carbonate** A compound often used in water softening operations, also called "soda ash." Chemical formula is $Na_2CO_3$.

**sodium cycle exchange** Ion exchange water softening process using a zeolite resin bed.

**sodium hexametaphosphate** A water-soluble polyphosphate commonly used as a sequestering or dispersing agent.

**sodium hydroxide** Caustic soda. Chemical formula is NaOH.

**sodium hypochlorite** A liquid chlorine solution frequently used as a water or wastewater disinfectant. Chemical formula is $Na(OCl)_2$.

**sodium metabisulfite** A crystalline form of sulfur dioxide used to remove chlorine. Chemical formula is $Na_2S_2O_5$.

**sodium sulfite** An oxygen scavenger used in boiler and cooling water systems. Chemical formula is $Na_2SO_3$.

**softening** Treatment process that involves the removal of calcium and magnesium ions from water.

**soil vapor extraction (SVE)** Technique to remove volatile organic compounds and promote bioremediation of compounds in unsaturated soils.

*Sokalan* Antiscalant for desalination of seawater by BASF.

**sol** Colloidal dispersion of solids in liquid.

**solar constant** The rate at which the sun's radiant energy is received per unit area on a horizontal surface at the top of the earth's atmosphere.

**solar pond** Pond used to accomplish evaporation via direct solar heating.

**solar still** A distillation device that utilizes solar energy.

**solid bowl centrifuge** Continuous operation centrifuge consisting of a cylindrical, tapered bowl and an internal helical scroll, both revolving at slightly different speeds, to separate solids from water by means of centrifugal force.

*Solidex*   Screw press system by Bepex Corp.

**solidification**   A process where materials are added to waste to produce a solid.

**solids balance**   A mathematical representation of a treatment system that defines the amount of solids entering and exiting each unit treatment process.

**solids contact clarifier**   A clarifier in which liquid passes upward though a solids blanket and discharges at or near the surface.

**solids retention time (SRT)**   The mass of solids in a vessel (kg) divided by the solids removed (L/day).

*Solidur*   Ultrahigh-molecular-weight polyethylene components for chain and flight sludge collectors by Solidur Plastics Co.

**solid waste**   Garbage, refuse, sludge, and other discarded material resulting from community activities or commercial or industrial operations.

*Solo*™   Controllerless downwell cleanup pumps by QED Environmental Systems.

**solubility**   The amount of a substance that can dissolve in a solution under a given set of conditions.

**solubility product**   The equilibrium constant that describes the reaction by which a precipitate dissolves in pure water to form its constituent ions.

**soluble**   Capable of being dissolved in a fluid.

*Solu Comp*®   Water measurement instrument by Rosemount Analytical, Inc.

**solum**   The upper layers of soil including the top soil (A-horizon) and intermediate soil (B-horizon).

**solute**   A substance dissolved in a fluid.

**solution**   A liquid that contains dissolved solute.

**solvent**   Liquid capable of dissolving or dispersing one or more substances.

**solvent extraction**   The process of selectively extracting a liquid constituent from a wastewater using an organic solvent. Also called "liquid-liquid extraction."

*Solvo Salveger*   Vacuum-assisted distillation process by Hoyt Corp., Westport Environmental Systems.

*Som-A-System*   Sludge dewatering equipment by Somat Corp.

**sone**   A subjectively determined unit of the loudness of a sound.

*Sonix 100*   Tank-mounted chlorinator by Wallace & Tiernan, Inc.

*Sonozaire*™   Electronic odor control device by Howe-Baker Engineers, Inc.

**soot**   Carbon dust formed by incomplete combustion.

**sorbent**   A solid material used to concentrate dissolved solids.

*Sorb-N-C*   Packaged stack gas sorbent by Church & Dwight, Inc.

*Sorbond*   Sludge stabilizer by Colloid Environmental Technologies Co.

**sorption**   The concentration of dissolved solids through absorption or adsorption on a solid.

*SOS*   Stormwater overflow screens by John Meunier, Inc.

**SOUR**   See "specific oxygen uptake rate."

**sour environment**   Environment containing significant amounts of hydrogen sulfide.

**SOx**   See "sulfur oxides."

**Soxhlet extraction method**   An extraction process using trichlorofluoroethane to determine the oil and grease content in a liquid.

*Spaans*   Screw pump and screw conveyor products offered by Asdor.

**sparingly soluble compounds**  Term used to describe compounds with solubility ranges from near zero to a few thousand milligrams per liter.

*Sparjair*  Package wastewater treatment plant by Walker Process Equipment Co.

*Sparjer*  Aeration products by Walker Process Equipment Co.

*SparjLift*  One- or two-level air injection pump by Walker Process Equipment Co.

**SPCC**  Spill Prevention Control and Countermeasures.

**specific conductance**  The reciprocal of specific resistance, usually stated in micromhos per centimeter.

**specific gravity**  The ratio of the density of a substance to the density of water.

**specific oxygen uptake rate (SOUR)**  An indicator of a sludge's odor causing potential, by aerating a sludge sample and measuring the rate of oxygen depletion.

**specific resistance**  A measure of total ionized solids concentration determined by the resistance of a $1 \text{ cm}^3$ of water to the passage of electricity under standard conditions.

**spectrophotometer**  An instrument for measuring the amount of electromagnetic radiation absorbed by a sample, as a function of wavelength.

*Spectrum™*  Aeration mixing systems by Environmental Dynamics, Inc.

**spent regenerant**  Wastes from the regeneration of an ion exchange system.

**Sphaerotilus**  A filamentous bacterium that commonly causes sludge bulking in activated sludge wastewater treatment plants.

*Spher-Flo*  Single-stage sewage pumps by Aurora Pumps.

*Spiracone®*  Conical tank upflow clarifier by General Filter Co.

*Spiraflo*    Peripheral feed clarifier by Lakeside Equipment Corp.

*Spirafloc*    Peripheral feed clarifier with flocculation zone in outer raceway by Lakeside Equipment Corp.

*Spiragester*    Combination digester and clarifier by Lakeside Equipment Corp.

*Spiragrit*    Grit removal system by Lakeside Equipment Corp.

*Spiralift*™    Screw pump by Zimpro Environmental, Inc.

*Spiralklean*    Screenings washer by Parkson Corp.

*Spiral Scoop*    Dissolved-air flotation skimming device by Krofta Engineering Corp.

*Spiraltek*™    Dry sump filters by Osmonics, Inc.

*Spira-Pac*    Package digester/clarifier combination by Lakeside Equipment Corp.

*Spiratex*™    Point-of-use filtration system by Osmonics, Inc.

*Spirathickener*    Peripheral feed gravity thickener by Lakeside Equipment Corp.

*Spiratrex*    Ultrafiltration membrane by Osmonics, Inc.

*Spira-Twin Spiragester*    Clarifier/digester unit for primary and secondary sedimentation of trickling filter effluents by Lakeside Equipment Corp.

*Spiravac*    Peripheral feed clarifier with vacuum-assisted sludge removal by Lakeside Equipment Corp.

*Spirolift*®    Screw-type vertical conveyor by Spirac AB (Europe) and JDV Equipment Corp. (U.S.).

*Spiropac*®    Screw-type compactor by Spirac AB (Europe) and JDV Equipment Corp. (U.S.).

*Spiropress*®    Screw-type solids dewatering press by Spirac AB (Europe) and JDV Equipment Corp. (U.S.).

*Spirosand*®    Shaftless grit classifier by Andritz-Ruthner, Inc.

*Spirovortex*    Activated sludge treatment system including tertiary filtration by Dorr-Oliver, Inc.

*Split-ClarAtor*™    Secondary clarifier by Aero-Mod, Inc.

**splitter box**    A chamber that equally divides incoming flow into two or more streams.

**sponge ball cleaning**    The use of flexible sponge balls added to a recirculating liquid to scour scale or other deposits that form on the inside of condenser or heat exchanger tubes.

**spore**    A reproductive cell or seed of a microbe, often dormant or environmentally resistant.

**spray dryer**    Sludge drying device utilizing centrifugal force to atomize sludge into fine particles and spray them into the top of a drying chamber.

*Spray-Film*®    Vapor compression distillation unit by Aqua-Chem, Inc.

**spray irrigation**    The spreading of treated wastewater on agricultural land by spraying.

*Spraymaster*®    Low-headroom packaged deaerator by Cleaver-Brooks.

**spring**    A natural flow of water from the ground.

**spring turnover**    See "turnover."

**spring water**    Bottled water collected from an underground formation from which the water flows naturally from the surface, or a bore hole that taps the spring and is located near where the spring emerges.

*Sprint*™    Submersible pump product line by Crane Pumps & Systems.

*Sprout-Bauer*    Former name of Andritz-Ruthner, Inc., screening equipment product line.

**SQG**    See "small quantity generator."

*Squarex*    Circular sludge collector mechanism with pivoted corner extensions for square basins by Dorr-Oliver, Inc.

**SR-7™**   Ion exchange resins by Sybron Chemicals, Inc.

**SRT**   See "solids retention time."

**SS**   See "suspended solids."

**SSO**   See "sanitary sewer overflow."

**SSPC**   Steel Structures Painting Council.

**SST**   See "stainless steel."

**SSV**   See "settled sludge volume."

*Stabilaire*   Package contact stabilization treatment plant by FMC Corp., MHS Division.

**stabilization pond**   A large shallow basin used for wastewater treatment by natural processes involving the use of algae and bacteria to accomplish biological oxidation of organic matter.

**stack**   Common term for the key element of an electrodialysis unit consisting of multiple membrane cells with electrodes on both ends.

**stage**   One of several units of a flash evaporator, each of which operates at a successively lower pressure.

**stagnant**   Motionless, not flowing in a current or stream.

*Stahlermatic®*   Rotating biological contact aerator by IBERO Anlagentechnik, GmbH.

**stainless steel**   Corrosion-resistant steel containing a minimum of 12% chromium as the principal alloying element.

*STAKfilter*   High-volume vertical pressure screen filter by Everfilt Corp.

*Stak-Tracker™*   Continuous emissions monitoring system by Reuter-Stokes.

**Standard Industrial Classification (SIC)**   A U.S. government numbering system used to categorize industrial facilities.

**Standard Methods**   "Standard Methods for the Examination of Water and Wastewater," published jointly by the American Public Health Association, American

Water Works Association, and Water Environment Federation.

**standard seawater** A widely accepted "standard" total dissolved solids concentration of approximately 36,000 mg/L, considered to be typical of most seawaters.

**standard solution** A solution whose strength or reacting value per unit volume is known.

**standpipe** A water storage reservoir having a cylindrical structure of uniform diameter with a height greater than its diameter.

*Stanley Compo-Cast* Former manufacturer of nonmetallic sludge collector wear shoes now made by Trusty Cook, Inc.

**stapling** The entanglement of stringy or fibrous debris on a mesh or bar rack.

*Star*™ Anaerobic package plants by EnviroSystems Supply, Inc.

*Star Filter*® Filter press by Star Systems, Inc.

*StarScreen* In-channel fine screen by OVRC Environmental (U.S.) and Sernagiotto (Italy).

*Sta-Sieve* Static fine screen by SWECO Engineering Co.

**State Implementation Plan (SIP)** Plans to implement air quality standards required of each state under the Clean Air Act.

**static head** The vertical distance between a fluid's supply surface level and free discharge level.

*Static Mixaerator*™ Static mixing aerators by Ralph B. Carter Co.

**static mixer** Motionless device consisting of fixed baffles incorporated into channels or pipelines to create turbulence and mix additives added upstream.

**static pile composting** Sludge composting method where piles of sludge are aerated to eliminate the need for remixing.

**static screen**   Fine screen using a stationary, inclined screen deck that acts as a sieve to remove solids from liquids.

**static tube diffuser**   A coarse-bubble diffuser consisting of a vertically oriented cylinder with internal baffles to promote air/water mixing.

**stator**   The stationary member of an electric motor or generator.

*Stato-Screen*   Static screen by Vulcan Industries, Inc.

*Stauffer Chemical*   Former name of Rhone-Poulenc Basic Chemical Co.

**steady state**   An equilibrium condition that exists in a system.

*Steady Stream*   Turbidimeter by Great Lakes Instruments, Inc.

**steam chest**   The steam chamber adjacent to a heat exchanger tubesheet.

**steam stripper**   The process of removing volatile and semivolatile contaminants from liquid where steam and liquid are passed countercurrently through a packed tower.

**STEL**   See "short-term exposure limit."

*Stengel Baffle*   Inlet baffle for rectangular sludge collector by Zimpro Environmental, Inc.

**stenothermophiles**   Bacteria that grow best at temperatures above 60°C.

**step aeration**   Variation of the activated sludge process where settled wastewater is introduced at several points in the aeration tank to equalize the food-to-microorganism ratio.

*Stepaire*   Circular package step aeration wastewater treatment plant by FMC Corp., MHS Division.

*Step Screen*   In-channel fine screen by Hycor Corp. (U.S.) and Hydropress Wallender (Sweden).

**sterile**   Free from bacteria or other microorganisms.

**sterilization**   The destruction or removal of all living organisms within a system.

*SternPAC*™   Polyaluminum coagulant by Sternson Ltd.

**Stiff & Davis Index**   Index used to determine the saturation point of calcium carbonate in seawater or other highly saline water.

**STIG**   Steam-injected gas turbine.

**still**   Apparatus used in distillation.

**stilling well**   A tube or chamber used to dampen waves or surges in a large body of water; usually used for purposes of water level measurement.

*Sti-P³*®   Steel tank standard for double-wall underground tank with cathodic protection by Steel Tank Institute.

**stock solution**   A concentrated chemical solution, often for use as a reagent.

**stoichiometric**   The ratio of chemical substances reacting in water that corresponds to their combining weights in the theoretical chemical reaction.

**stoker**   A mechanical device that feeds solid fuel to a furnace.

**Stokes law**   The settling velocity of a particle based on its density and size.

**stop log**   A removable wooden, steel, or concrete bulkhead that fits in vertical grooves in a channel to stop water flow.

**storage lagoon**   A lagoon constructed with a sealed bottom used to store and collect solids.

*Storm King*™   Vortex-type separation system by H.I.L. Technology, Inc.

**storm sewer**   Collection system of underground piping used to remove water resulting from precipitation runoff.

*StormTreat*™  Stormwater collection and treatment system by StormTreat Systems.

**stormwater**  Water resulting from precipitation runoff.

**STP**  Standard temperature and pressure.

**STP**  Sewage treatment plant.

*Straightline*®  Group of wastewater treatment products by FMC Corp., MHS Division.

**strainer**  A device that retains solids, but allows liquids to flow through.

*Strain-O-Matic*  Self-cleaning strainer by Hayward Industrial Products, Inc.

*StrainPress*  Compactor formerly offered by Parkson Corp.

*Strantrol*  Chemical controller for disinfection systems by Stranco, Inc.

*Strata Clear*  Water flotation separator by Smith & Loveless, Inc.

*StrataMix*  Fine-bubble diffuser by Wilfley Weber, Inc.

*Strata-Sand*  Continuously backwashed gravity sand filter by F.B. Leopold Company, Inc. (U.S.) and Simon-Hartley (U.K.).

*Stratavap*  Evaporator system by Licon, Inc.

**stratosphere**  The level of the atmosphere containing most of the earth's ozone that lies between the troposphere and the mesosphere.

**stray current corrosion**  Electrochemical corrosion caused by stray currents leaking from an electrical installation.

**streaming current**  The net ionic and colloidal surface charges of suspended solid particles in solution.

**streaming current detector (SCD)**  Measuring device used to detect and monitor the net electrical charge of particles in solution after coagulants have been added.

*Streamline*   Flow proportional sampler by American Sigma, Inc.

*Streamline*   Rectangular chain and flight collector by Purestream, Inc.

*Stream Saver*™   Automated spill control system by ILC Dover.

**streptococcus**   A genus of bacteria that includes some of the most common human pathogens.

**stress corrosion cracking**   Formation of cracks caused by the action of a corrosive medium in combination with tensile stress.

*Stress-Key*   Precast concrete packaged sewage treatment plants by Marolf, Inc.

**stress relieving**   Heat treatment carried out to reduce internal stresses in steel.

**strip mining**   A method of mining where surface soil and strata are removed to gain access to the mineral deposits.

**stripper**   Device used to remove volatile and semivolatile contaminants from water.

*Stripper*®   Multistaged diffused-bubble aeration system by Lowry Aeration Systems, Inc.

*Stripperator*   Treatment unit for hydrocarbon contaminated water by Ejector Systems, Inc.

**strong acid**   An acid that approaches 100% ionization in dilute solutions.

*Stuart-Carter*   Walking beam flocculator by Ralph B. Carter Co.

**subdrainage**   The control and removal of excess groundwater, usually by means of pipe drains that intercept seepage and/or lower the water table.

**subituminous coal**   A grade of coal with a heat content higher than that of lignite but lower than bituminous.

**sublimation**   The process of changing a gas to a solid, or a solid to a gas, without going through the liquid phase.

**submerged tube evaporator**   An evaporator in which steam enters a tube bundle submerged in the fluid to be boiled.

**subnatant**   Liquid remaining beneath the surface of floating solids.

*Suboscreen®*   In-channel rotary fine screen by Andritz-Ruthner, Inc. (Western Hemisphere) and Contra-Shear Engineering, Ltd.

*Subrotor*   Progressing cavity pump by Ingersoll-Dresser Pump (U.S.) and H2O Waste-Tec (U.K.).

**subsidence**   The lowering of the natural land surface due to a reduction of the fluid pressure, removal of underlying supporting material, compaction due to wetting, or added loads on the land surface.

**subsonic flow**   Liquid flow at a speed less than the speed of sound in the fluid.

**substrate**   The organic matter or nutrients that are used as food substances during biological wastewater treatment.

**sulfamic acid**   Acid often used as a cleaning agent. Chemical formula is $HSO_3NH_2$.

**sulfate reducing bacteria**   Bacteria capable of reducing sulfate or other forms of oxidized sulfur to hydrogen sulfide.

*SulfaTreat®*   Hydrogen sulfide removal process by SulfaTreat Co.

*Sulfaver*   Reagent chemical used to determine phosphates concentration in water by Hach Company.

*Sulf Control®*   Sulfide inhibitor for prevention of hydrogen sulfide formation by NuTech Environmental Corp.

*Sulfex®*   Sulfide precipitation process for heavy metal removal by U.S. Filter Corp.

**sulfonator**   Device used to inject and meter sulfur dioxide to dechlorinate water.

**sulfuric acid** A toxic corrosive acid capable of dissolving most metals. Chemical formula is $H_2SO_4$.

**sulfur oxides (SOx)** Air contaminants resulting from the combustion of fuels containing sulfur in the presence of oxygen.

**sullage** See "gray water."

*Sulzer* Screenings grinder by Dorr-Oliver, Inc.

*Sumigate®* Rubber dam for combined sewer overflow applications by Rodney Hunt Co. and Sumitomo Electric Industries, Inc.

**sump** A pit or reservoir that serves to collect water or wastewater for subsequent removal from the system.

*Sumpaire™* Self-aspirating jet aerator by Framco Environmental Technologies.

*Sump-Gard®* Vertical centrifugal pump by Vanton Pump & Equipment Corp.

*Suparator®* Oil/water separator by Lemacon Techniek B.V.

**super austenitic stainless steel** Stainless steel alloyed with more than 4% molybdenum.

*Super Blend™* Cellulose acetate reverse osmosis membranes by TriSep Corp.

*Superblock II®* Filter underdrain with water and air/water backwash capabilities by F.B. Leopold Co.

*Super Blue®* Bolted steel storage tank by U.S. Filter Corp.

*Super-Cel®* Diatomaceous earth filter media by Celite Corp.

*Supercell* Dissolved-air flotation unit by Krofta Engineering Corp.

**supercool** To cool a substance in its liquid state at a temperature below its freezing point.

**supercritical water oxidation (SCWO)**   High-temperature/pressure wastewater treatment process where organic material is oxidized at temperatures above a fluid's critical point.

*Super Detox*™   Encapsulation process to chemically stabilize furnace dust and other heavy metal residues by Conversion Systems, Inc.

*Super Dome*™   Ceramic dome diffuser by Ferro Corp.

*Superdraw*   Supernatant withdrawal unit by Walker Process Equipment Co.

*Superfloc Excel*   High-charge cationic flocculant by Cytec Industries, Inc.

**Superfund**   U.S. federal law authorizing identification and remediation of unsupervised hazardous waste sites. Also "CERCLA."

**Superfund Amendments and Reauthorization Act**
1986 U.S. law passed to reauthorize and expand the Comprehensive Environmental Response, Compensation and Liability Act, and require public disclosure of chemical release information and development of an emergency response plan.

**Superfund Innovative Technology Evaluation (SITE)**   EPA-supported research, development, and demonstration projects designed to develop new remediation technologies.

**superheat**   The sensible heat in a gas above the amount needed to maintain the gas phase.

**superheated steam**   Steam with additional heat added after vaporization, increasing its temperature and energy.

*Supermal*   Pearlitic malleable iron chain material by Jeffrey Chain Co.

**supernatant** Liquid above the settled sludge layer in a sedimentation basin.

*Superpulsator®* Solids contact clarifier utilizing inclined plates and intermittent pulsing to expand the sludge blanket by Infilco Degremont, Inc.

*Super Shredder®* In-line macerator by Franklin Miller.

*Super Sieve Screen* Sieve screen by Sizetec, Inc.

*Superslant™* Inclined plates for clarification by Filtronics, Inc.

*Superstill* Former name of VaPure vapor compression still by Mueller.

*Superthickener* Large diameter center pier gravity thickener by Dorr-Oliver, Inc.

**support gravel** Layers of graded gravel between underdrain openings and filter media to prevent media from leaking into the underdrain.

*Supracell* Dissolved-air flotation unit by Krofta Engineering Corp.

*Suprex™* Mixed bed type condensate polisher by Graver Co.

*SURF®* Two-stage contact clarifier/filter by General Filter Co.

**surface aerator** Mechanical aeration device consisting of a partially submerged impeller, attached to a motor, and mounted on floats or a fixed structure.

**surface condenser** A condenser, usually of the shell-and-tube design, that provides a suitable heat transfer surface area for condensing to occur, where cooling water and process fluid remain separated.

**surface impoundment** A natural topographic depression, manmade excavation, or diked area formed primarily of earthen materials, to hold an accumulation of liquid wastes or waste containing free liquids.

**surface loading rate** A criteria used for design of sedimentation tanks expressed as flow per day per unit of basin surface area. Also called "overflow rate."

**surface mining** See "strip mining."

**Surface Mining Control and Reclamation Act (SMCRA)** 1977 U.S. law that set performance standards for environmental protection to be met at most surface mining operations for coal.

*Surface Scatter* On-line turbidimeter by Hach Company.

**surface tension** The force acting on a liquid surface that results in a minimum liquid surface area. Produced by the unbalanced inward pull exerted on the layer of surface molecules by molecules below the liquid surface.

**surface wash** An auxiliary high-pressure water spray system used to agitate and wash the surface of granular media filters.

**surface water** Water from sources open to the atmosphere including lakes, reservoirs, rivers, and streams.

*Surfact* Process to upgrade activated sludge system with air-driven rotating biological contactors by Envirex, Inc.

**surfactant** A surface active agent such as a detergent that, when mixed with water, generally increases its cleaning ability, solubility, and penetration, while reducing its surface tension.

*Surfaer* Slow-speed surface aerator by Aerators, Inc.

*Surfpac*™ Vertical-type polyvinyl chloride trickling filter media by American Surfpac Corp.

**surge-flow irrigation** A surface irrigation technique that involves intermittent application of irrigation water to a field for 50% of an irrigation cycle and diversion to another area for the remainder of the cycle.

**suspended growth process** Biological wastewater treatment process where the microbes and substrate are maintained in suspension within the liquid.

**suspended solids (SS)**   Solids captured by filtration through a glass wool mat or 0.45-micron filter membrane.

**suspension**   A system in which very small particles are uniformly dispersed in a liquid or gaseous medium.

**suspensoid**   Colloidal dispersion of solids in liquid.

*Sutorbilt®*   Air blower by Gardner-Denver, a division of Cooper Industries.

**SVE**   See "soil vapor extraction."

**SVI**   See "sludge volume index."

*SVVS™*   Subsurface bioremediation system by Brown & Root Environmental.

**swamp gas**   See "marsh gas."

**sweep flocculation**   Coagulation/flocculation process using relatively large amounts of iron or metal salts to form voluminous floc particles used to trap smaller particles.

**sweeten**   To remove sulfur contaminants from petroleum products.

**sweet environment**   Environment containing no, or negligible amounts of, hydrogen sulfide.

**sweet water**   Brackish water that may be used for drinking even though it may not meet potable water standards.

**swelling**   Condition that exists as a result of intrusion of water into a particle.

*Swing-Flex™*   Check valve by Val-Matic, Corp.

*Swingfuser®*   Removable aeration header and drop pipe assembly by FMC Corp., MHS Division.

*Swingtherm®*   Regenerative catalytic oxidizer by MoDo-Chementics.

*SwingUp*   Removable aeration header and drop pipe assembly by Aerators, Inc.

*Swingwirl*   Vortex flowmeter by Endress+Hauser.

*Swirl-Flo*™   Solids/liquid separation process by H.I.L. Technology, Inc.

*SwirlMix*   Complete mix package wastewater treatment plant by Walker Process Equipment Co.

*Swiss Combi System*   Sludge drying and pelletization system by Wheelabrator Clean Water Systems, Inc., Bio Gro Division.

**SWRO**   Seawater reverse osmosis.

**SWTR**   Surface Water Treatment Rule.

**symbiotic**   A relationship between organisms of different species that benefits both members of the relationship such that neither could carry out certain activities alone.

**synergy**   The combined action of two agents that results in a reaction greater than the sum of the individual agents acting alone.

**synfuels**   Liquid or gaseous fuels produced from coal, lignite, or other solid carbon sources.

*Syphonid*   Recovery device for sinker contaminates by R.E. Wright Associates, Inc.

*System-3*   Oily water treatment system by Megator.

# T

**tailings**    Residue from the separation of useful values from ore.

*Tait-Andritz*    Former name of Andritz-Ruthner, Inc.

**tangential screen**    See "static screen."

**tannin**    Colored compounds that form when plant matter degrades in water.

**tapered aeration**    Variation of the activated sludge process where the amount of air supplied in an aeration basin is tapered to match the demand exerted by the microbes.

**tapered flocculation**    A flocculation process utilizing multiple compartments and a gradually increasing velocity gradient.

**tapeworm**    A parasitic flatworm that can live in the digestive tract or liver of vertebrates.

**TAPPI**    Technical Association of the Pulp and Paper Industry.

**tare**    The empty weight of a vessel or container determined by deducting the weight of the contents from the total weight of the full load.

*Targa*    Vapor compression water desalination unit by Mechanical Equipment Co., Inc.

*Taskmaster*®    Screenings grinder by Franklin Miller.

*Taulman/Weiss*    In-vessel composting system by Taulman Company.

**TBT**    See "top brine temperature."

**TCD**    Thermocompression distillation.

**TCDD**    Dioxin (tetrachlorodibenzo-*p*-dioxin).

**TCE**   See "trichloroethylene."

*TCF®*   Horizontal tube cartridge filter by Ropur AG.

**TCLP**   See "toxicity characteristic leaching procedure."

**TCR**   See "Total Coliform Rule."

**TDH**   See "total dynamic head."

**TDS**   See "total dissolved solids."

*Teacup™*   Grit removal system by Eutek Systems.

**TEC**   Tonka Equipment Company.

*TechXtract™*   Chemical decontamination process by EET, Inc.

*TecTank*   Bolted steel storage tank by Peabody TecTank.

*Tecweigh®*   Volumetric feeder by Tecnetics Industries, Inc.

*TEES®*   High-pressure catalytic process by Onsite-Offsite, Inc.

**TEFC**   See "totally enclosed fan cooled."

*Tekleen™*   Filter screen by Automatic Filters, Inc.

**telemetry**   The process of transmitting measured data by radio to a distant station.

*TeleTote™*   Open-channel electromagnetic flowmeter by Marsh-McBirney, Inc.

**TEM**   See "transmission electron microscope."

**TEMA**   Tubular Equipment Manufacturers Association.

**temperature inversion**   See "inversion."

**temporary hardness**   Hardness associated with bicarbonates of calcium and magnesium that precipitate upon boiling.

**temporary threshold shift (TTS)**   A temporary reduction in hearing ability that results from noise overexposure.

**tensile strength**   The maximum tensile load per square unit of cross-section that a material is able to withstand.

**Ten States Standards**   Common name for "Recommended Standards for Sewage Works by the Great

Lakes–Upper Mississippi River Board of State Sanitary Engineers."

*Tenten*   Continuously backwashed gravity sand filter by F.B. Leopold Company, Inc. (U.S.) and Simon-Hartley (U.K.).

**TENV**   See "totally enclosed nonventilated."

**teratogenic**   A chemical or agent with properties that cause disfiguring by affecting the genetic characteristics of an organism.

**terminal headloss**   The headloss at the end of a filter run cycle signifying that the filter bed is filled with solids.

**terminal settling velocity**   The maximum sedimentation rate of an unhindered suspended particle.

*Terra-Gator*   Pressurized waste sludge injection system by Ag-Chem Equipment Co.

**tertiary filtration**   The use of a granular media filter to improve secondary effluent quality.

**tertiary treatment**   The use of physical, chemical, or biological means to improve secondary effluent quality.

**tetrachloroethane**   A chlorinated hydrocarbon used as an industrial cleaner or solvent.

*Texas Star*   Circular membrane diffuser by Aeration Research Company.

**TF/AS**   See "trickling filter-activated sludge."

*TFC®*   Reverse osmosis membrane by Fluid Systems Corp.

*TFCL®*   Thin-film composite reverse osmosis membrane by Fluid Systems Corp.

*TFCS™*   Reverse osmosis elements by Fluid Systems Corp.

*TFM®*   Reverse osmosis membrane by Desalination Systems, Inc.

**TF/SC**   Trickling filter-solids contact.

**TGNMO** Total gaseous nonmethane organics.

**theoretical oxygen demand (ThOD)** Determination of organic matter in a water or wastewater calculated on the basis of the chemical formula of the constituents.

*ThermaGrid™* Regenerative thermal oxidizer by United McGill Corp.

**thermal oxidizer** An emissions control device that utilizes heat to accomplish volatile organic compound oxidation.

**thermal pollution** The discharge of heated substances into the environment resulting in an undesired increase in ambient temperature, above that resulting from natural solar radiation.

**thermal value** See "heat value."

*ThermoBlender™* Sludge and quicklime blending unit by RDP Company.

**thermocline** The middle layer in a stratified lake that results from varying water densities.

**thermocompressor** A steam ejector that uses high-pressure steam to increase the pressure of a lower pressure steam.

**thermophiles** Bacteria that grow best at temperatures between 45 and 60°C.

**thermophilic digestion** Sludge digestion within a thermophilic range of approximately 45 to 60°C.

*Thermo-Sludge Dewatering™* Float-type sludge thickening process by Dontech, Inc.

*Thermox®* Flue gas analyzer by Ametek Inc., Process & Analytical Division.

**thickener** A tank, vessel, or apparatus that is used to reduce the proportion of water in a slurry or sludge.

**thickening** A procedure used to increase the solids content of sludge by removing a portion of the liquid.

**thin-film evaporator**  Evaporator where liquid flows or is sprayed over heat transfer surfaces, usually tubes, in a thin, turbulent film.

**thiols**  See "mercaptans."

**thixotropy**  The time-dependent property of some emulsions and sludges to change rheological and physical characteristics when left at rest.

**THM**  See "trihalomethane."

**THMFP**  Trihalomethane formation potential.

**ThOD**  See "theoretical oxygen demand."

*3DP*™  Three-belt dewatering filter by Eimco Process Equipment Co.

**threshold dose**  The minimum dose of a substance to produce an effect.

**threshold limit value (TLV)**  The maximal allowable workplace air level for a chemical.

**threshold odor number (TON)**  The number of dilutions of odor-free water required to eliminate the odor in a water sample.

*Thru Clean*  Back-cleaned mechanical bar screen by FMC Corp., MHS Division.

**tide**  The periodic rising and falling of the sea that results from the gravitational attraction of the moon.

*Tideflex*™  Diffuser check valve by Red Valve Co., Inc.

*Tile*  Ceramic tile filter underdrain by Roberts Filter Manufacturing Co.

**tile field**  A system used for subsurface discharge of treated wastewater effluents using open-jointed tile placed on gravel fill.

**tilted plate separator**  Oil separation device utilizing inclined plates to separate free nonemulsified oil and water based on their density difference.

**TIN**  Total inorganic nitrogen.

**tinajero** A home water filtration device where water is collected in a clay pot after filtration through a porous stone.

**tines** The teeth or prongs of a bar screen cleaning rake.

**tipping** The dumping of the contents of a waste truck at a waste disposal facility.

**tipping fee** The fee charged to dispose of solid waste at a sanitary landfill.

*Tipping Scum Weir* Pivoting scum weir by F.B. Leopold Co., Inc.

*Titeseal* Copolymer fluid control gate by Plasti-Fab, Inc.

**titration** A method of determining the concentration of a dissolved substance in terms of the smallest amount of a reagent required to bring about a given effect in reaction with a known volume of test solution.

*Titraver* Chemicals used in analysis of water hardness by Hach Company.

**TKN** See "total Kjeldahl nitrogen."

*TLC*™ Thin-layer composite reverse osmosis membrane by Osmonics.

**TLV** See "threshold limit value."

**TMDL** Total maximum daily load.

**T&O** Taste and odor.

**TOA** Trace organic analysis.

**TOC** See "total organic carbon."

**TOD** See "total oxygen demand."

*Tolhurst* Laboratory basket centrifuge by Ketema, Inc.

**TON** See "threshold odor number."

**ton** A unit of weight equal to 2000 lb or 907.2 kg. Also called short ton.

**ton container** A 1-ton chlorine storage container.

*TonkaFlo* Pumps for RO/UF/DI applications by Osmonics, Inc.

**tonne** See "metric ton."

**top brine temperature** The maximum temperature of the fluid being evaporated in an evaporator system.

*Tornado®* Aspirating-type surface aerator by Aeromix Systems, Inc.

*Torpedo™* Clarifier foam and grease removal unit by United Industries, Inc.

*Torpedo Filter* Floating microfilter by BTG, Inc.

*Torque-Flow* Grit handling vortex slurry pump by Envirotech Company.

*TorusDisc* Sludge dryer/cooler by Bepex Corp.

*Torvex®* Catalytic oxidation system by CSM Environmental Systems, Inc.

**Total Coliform Rule (TCR)** An EPA rule regulating known pathogens in drinking water.

**total dissolved solids (TDS)** The sum of all volatile and nonvolatile solids dissolved in a water or wastewater.

**total dynamic head (TDH)** The total energy that a pump must impart to water to move it from one point to another, measured as the difference in height between the free water surface level on the discharge and suction sides of a pump.

**total Kjeldahl nitrogen (TKN)** Organic nitrogen plus ammonia nitrogen.

**totally enclosed fan cooled (TEFC)** Designation for motor enclosure that is not airtight, but does not allow free exchange of air between the inside and outside of the motor case. Exterior cooling is provided by an integral external fan.

**totally enclosed nonventilated (TENV)** Designation for motor enclosure that is not airtight, but constructed so as to prevent free exchange of air between the inside and outside of the motor case.

**total organic carbon (TOC)**   Measurement of organic matter in a water or wastewater that can be oxidized in a high-temperature furnace.

**total oxygen demand (TOD)**   Measurement of organic matter in a water or wastewater that can be converted to stable endproducts in a platinum-catalyzed combustion chamber.

**total solids (TS)**   (1) The sum of dissolved and suspended solids in a water or wastewater. (2) Matter remaining as residue upon evaporation at 103 to 105°C.

**total suspended particulates (TSP)**   The concentration of all airborne particulate matter, usually expressed as micrograms of particulate per cubic meter of sampled air.

**total suspended solids (TSS)**   The measure of particulate matter suspended in a sample of water or wastewater. After filtering a sample of a known volume, the filter is dried and weighed to determine the residue retained.

**total trihalomethane (TTHM)**   The sum of the concentration of trihalomethane compounds rounded to two significant figures.

*Toveko*®   Continuously operating sand filter by Krüger, Inc.

*Tow-Bro*®   Suction-type sludge removal system for circular clarifiers by Envirex, Inc.

*Tower Biology*®   Biological wastewater treatment system by Biwater Treatment Ltd., licensed from Bayer AG.

*Towerbrom*®   Bromine-based biocide by Calgon Corp.

*Towermaster*   Pressurized sand filter for cooling tower applications by Serck Baker, Inc.

*Tower Press*   Belt filter press by Roediger Pittsburgh, Inc.

*ToxAlarm*™   On-line toxicity monitor by Anatel Corp.

**toxic**   Capable of causing an adverse effect on biological tissue following physical contact or absorption.

**toxicant**   A substance that is toxic to another organism.

**toxicity**   The property of being poisonous, or causing an adverse effect on a living organism.

**toxicity characteristic leaching procedure (TCLP)**   Method used to determine the amount of a hazardous substance that will leach from a solid when the solid is subjected to water.

**toxicology**   The study of adverse effects of chemicals on living organisms.

**Toxic Substances Control Act (TSCA)**   1976 U.S. law authorizing the EPA to collect information on chemical risks.

**toxic waste**   A waste that can produce injury upon contact with, or by accumulation in or on, the body of a living organism.

*Toxilog*   Portable single-gas detector by Biosystems, Inc.

**toxin**   A poisonous material that can cause damage to biological tissue following physical contact or absorption.

*TPC®*   Potassium permanganate by Technical Products Corp.

**trace elements**   (1) Any elements in a water or wastewater that are present in very low concentrations. (2) Elements present in minor amounts in the earth's crust.

**trace organics**   Organic matter present in water supplies in very low concentrations that originates from natural sources and the synthetic chemical industry.

*Trac Pump*   Sludge dredging system by H&H Pump and Dredge Co.

*Tractor Drive*   Circular clarifier drive unit by Walker Process Equipment Co.

*TracVac™*   Suction sludge removal mechanism by Eimco Process Equipment Co.

*TracWare™*   Particle counting system by Chemtrac Systems, Inc.

**transducer**   A device that receives energy from one system and retransmits it, often in another form, to another system.

*Transmax*®   Medium/coarse-bubble air diffuser by Enviroquip, Inc.

**transmission electron microscope (TEM)**   A microscope with a magnification range of 220 to 1,000,000× at a resolution of 2 Å, where illumination is provided by a beam of electrons that pass through the electron transparent specimen.

**transmissivity**   The rate at which water flows through an aquifer under a hydraulic gradient.

**transmutation**   Changing one element into another by changing the number of protons in the atom nucleus.

*Trans-Pak*®   Solid waste compacting/baling unit by Harris Waste Management Group, Inc.

**transpiration**   The loss of water from the leaves and stems of plants.

**transuranic wastes**   Radioactive wastes containing isotopes above uranium in the periodic table that are byproducts of fuel assembly, weapons fabrication, and reprocessing.

*Transvap*   Mobile wastewater reduction/recovery system by Licon, Inc.

*Trasar*®   Control and diagnostic technology by Nalco Chemical Co.

**trash**   Combustible waste including up to 10% plastic or rubber scraps from commercial and industrial sources.

*Trash Hog*®   Self-priming, solids handling pumps by ITT Marlow.

**trash rack**   Coarse intake screen usually consisting of vertical bars spaced to provide 25–75 mm openings.

**trash rake**   A mechanical screening device used to remove rough, debris, or trash from a trash rack.

*Travalift*™　Traversing sludge collecting and pumping mechanism for package treatment plants by FMC Corp., MHS Division.

*Traveco*　Traveling water screen by Eco Equipment.

**traveling bridge clarifier**　A rectangular clarifier with the sludge removal mechanism supported by a traversing bridge.

**traveling bridge filter**　A granular media filter with multiple compartments that can be individually cleaned by a moveable, bridge-mounted backwashing device without taking the entire filter out of service.

**traveling screen**　See "traveling water screen."

**traveling water screen (TWS)**　Automatically cleaned screening devices employing chain-mounted wire mesh panels to remove floating or suspended solids from a channel of water.

*Traypak*™　Deaerator trays by Graver Co.

**TRC**　Total residual chlorine.

**treatability study**　A study in which a waste is subjected to a treatment process to determine whether it is amenable to treatment and/or to determine the treatment efficiency or optimal process conditions for treatment.

**treatment, storage, and disposal (TSD)**　A description of a facility where hazardous waste is treated, stored, and/or disposed.

*Trebler*　Automatic sampler by Lakeside Equipment Corp.

*Trey Deaerator*™　Three-stage deaerator by U.S. Filter Corp.

**trial burn**　An incinerator test to demonstrate compliance of the unit with operating standards of the Resource Conservation and Recovery Act.

*Triangle Brand*®　Copper sulfate product by Phelps Dodge Refining Co.

*Triboflow*  Continuous particulate emission monitor by Auburn International.

*Tricanter*®  Centrifuge by Flottweg.

*Tricellorator*  Three-compartment dissolved-air flotation unit by Pollution Control Engineering, Inc.

*Tricep*™  Granular resin filtration process for the removal of iron and copper oxides with oil by Graver Co.

**trichloroethylene (TCE)**  A chlorinated hydrocarbon used as an industrial cleaner or solvent; it may cause organ damage or tumors in humans.

**trickling filter**  An aerobic, fixed-film wastewater treatment process where organic matter present in a wastewater is degraded as it is distributed over a biological filter bed.

**trickling filter-activated sludge (TF/AS)**  Process combining trickling filter wastewater treatment followed by an activated sludge process to satisfy some unique treatment requirement.

*Tricon*  Concrete water treatment plant with buoyant media flocculator/clarifier by Wheelabrator Engineered Systems, Inc., Microfloc Products.

*Tridair*  Dissolved-air flotation unit by Engineering Specialties, Inc.

*Trident*®  Modular water treatment plant with buoyant media flocculator/clarifier by Wheelabrator Engineered Systems, Inc., Microfloc Products.

**trihalomethane (THM)**  A disinfectant byproduct formed when chlorine reacts with organic compounds in water. These halogenated organics are named as derivatives of methane and include suspected carcinogens.

*Trimite*™  Package water treatment plant with buoyant media flocculator/clarifier by Wheelabrator Engineered Systems, Inc., Microfloc Products.

*Tri-NOx®*    Nitrogen oxides removal process by Tri-Mer
    Corp.

*Tri-Packs®*    Plastic packing media by Jaeger Products,
    Inc.

**triple point**    The condition at which the solid, liquid, and
    gaseous states of a substance coexist simultaneously in
    equilibrium.

*Triplex™*    Odor control, multistage air scrubber by Davis
    Water & Waste Industries, Inc.

*Trisep™*    Oil/solids/water separation filters by Graver Co.

*Triton*    Circular membrane diffuser by Aeration Research
    Company.

*Triton®*    Wedgewire screen lateral underdrain system by
    Wheelabrator Engineered Systems, Inc., Microfloc
    Products.

*Tritor*    Front-cleaned bar screen and grit removal device
    by FMC Corp.

*Triturator*    Screenings grinder by Envirex, Inc.

*TriZone™*    Compact water treatment plant by Wheelabra-
    tor Engineered Systems, Inc., Microfloc Products.

*Troll*    Submersible water and temperature level probe by
    In-Situ, Inc.

*Tromax™*    Trommel screen by Norkot Mfg. Co.

**trommel screen**    A cylindrical, rotating screen used to
    separate solid waste material according to size and den-
    sity.

**trophic level**    Any of the distinct feeding levels in a food
    chain.

**tropical rainforest**    The densely forested equatorial
    regions of the world that experience very high annual
    rainfall.

**troposphere**    The lowest level of the atmosphere, which
    extends to a height of between 9 and 16 km above the
    earth's surface.

*TroubleShooter*™   Portable containment unit for hazardous waste spills by Therma Fab, Inc.

**TRPH**   Total recoverable petroleum hydrocarbons.

**true color**   The color in water caused by the presence of humic or fulvic acids that result from the decomposition of organic matter.

*Tru-Grit*   Grit washing and separation system by Hycor Corp.

*Tru-Gritter*   Grit removal system by Vulcan Industries, Inc.

**trunnion**   A pivot or pin mounted on bearings used to rotate or tilt something.

*Tru-Test*   Automatic liquid sampler by FMC Corp., MHS Division.

**TS**   See "total solids."

**TSCA**   See "Toxic Substances Control Act."

**TSD**   See "treatment, storage, and disposal."

**TSE**   Treated sewage effluent.

**TSP**   See "total suspended particulates."

**TSS**   See "total suspended solids."

**TTHM**   See "total trihalomethane."

**TTS**   See "temporary threshold shift."

**TU**   Turbidity unit. See "nephelometric turbidity unit."

**tubercles**   Knob-like mounds of corrosion on pipe surfaces.

**tube settlers**   A series of parallel inclined tubes used to increase the settling efficiency of sedimentation basins.

**tubesheet**   Flat plate used to secure the ends of tubes in an evaporator, heat exchanger, or boiler.

*Tub Scrubber*   Single-stage dry chemical air scrubber by Purafil, Inc.

*Tuff-Span*   Modular tank covers by Composite Technology, Inc.

*Tulsion®*   High-temperature ion exchange resin by Thermax, Ltd.

*Tunnel Reactor®*   In-vessel composting system by Waste Solutions.

**turbidimeter**   Instrument used to measure water turbidity by detecting the intensity of light scattered at angles from a beam of light projected through a water sample.

**turbidity**   Suspended matter in water or wastewater that scatters or otherwise interferes with the passage of light through the water.

*Turbitrol®*   Division of the Taulman Company.

*Turbo™*   Reverse osmosis booster pump by Pump Engineering, Inc.

*TURBO™*   Floating, axial flow surface aerator by Aeration Industries Inc.

*TurboBlade™*   Low-speed mixer with variable pitch impeller blades by Eimco Process Equipment Co.

*TURBO-Dryer®*   Sludge dryer by Wyssmont Co., Inc.

*Turbofill™*   Random packed air stripper media by Diversified Remediation Controls, Inc.

*TurboFlow™*   Soil remediation exhausters and blowers by Invincible AirFlow Systems.

*Turbo-Scour*   Continuously cleaned package sand filters by Smith & Loveless, Inc.

*Turboshredder*   Cutter assembly for grinder pump by Homa Pumps, Inc.

*Turbostripper™*   Air stripper by Diversified Remediation Controls, Inc.

*Turbotron®*   Rotary lobe blower by Lamson Corp.

*Turbulator*   Rapid mixing unit by Walker Process Equipment Co.

**turbulent flow**   A flow situation in which the fluid moves in a random manner, with a Reynolds number usually greater than 4000.

**turndown**   A ratio expressing the maximum-to-minimum capacity of a process or device.

**turnover**   Seasonal (spring and fall) change that occurs in a lake's thermal gradients, resulting in the circulation of biological and chemical materials.

**TVC**   Thermal vapor compression. See "vapor compression evaporation."

**TVR**   Thermal vapor recompression. See "vapor compression evaporation."

**TWL**   Top water level.

**two-tray clarifier**   Space-saving wastewater clarifier arrangement where one longitudinal clarifier basin is located above another and both are operated in series.

**TWS**   See "traveling water screen."

*TxPro™*   Solids transmitter by BTG, Inc.

*Tync Aeration*   Former name of Aeration Research Company.

**typhoid**   Highly infectious disease of the gastrointestinal tract caused by waterborne bacteria.

*Tysul*   Hydrogen peroxide oxidant by E. I. du Pont de Nemours, Inc.

# U

**UAQI**   Uniform Air Quality Index.

**UBC**   Uniform Building Code.

**UF**   See "ultrafiltration."

**UHMW**   Ultrahigh-molecular-weight.

*UHR Filter*   Dual-media sand filter by Idreco USA, Ltd.

**UL**   Underwriters Laboratories.

*Ultima*   Nonmetallic sludge collector flight by Budd Co.

**ultimate BOD (BODu)**   The amount of oxygen required to completely satisfy the carbonaceous and nitrogenous biochemical oxygen demand.

**ultimate strength**   The stress, calculated on the maximum value of the force and the original area of the cross-section, that causes fracture of the material.

*Ultracept*®   Oil water separators by Jay R. Smith Mfg. Co., Environmental Products Group.

*UltraChem*™   Membrane for oily waste applications by Membrex, Inc.

**ultrafiltration (UF)**   A low-pressure membrane filtration process that separates solutes in the 20–1000 angstrom (up to 0.1 micron) size range.

*UltraFlex*®   LPDE containment liner by SLT North America, Inc.

*Ultraflow*   Programmable open-channel flow measurement device by Monitek Technologies, Inc.

*Ultramix*™   High-efficiency mixer motor by Lightin.

**ultrapure water**   Water with a specific resistance higher than 1 megohm/cm.

*UltraScrub*   Wet scrubber by Tri-Mer Corp.

*Ultrasep*  Ion exchange membrane for pressure-driven processes by Ionics, Inc.

**ultrasonic**  Acoustic waves with frequencies above 20 kHz, not audible to the human ear.

*UltraStrip*  Air stripper by GeoPure Continental.

*Ultratest*  Polyelectrolyte used to enhance liquid/solid separation by Ashland Chemical, Drew Division.

**ultraviolet (UV) light**  Light rays beyond the violet region in the visible spectrum, invisible to the human eye.

*Ultrex*  Tubular membrane diffuser by Aeration Research Company.

*Ultrion*  Cationic coagulants by Nalco Chemical Co.

*Ultrox*  Process using ultraviolet light and oxidants to destroy toxic organics by Ultrox International.

**unaccounted-for water**  The fraction of water that is fed into a water distribution system that is not registered by the customers' meters.

**unburned lime**  Another term for calcium carbonate.

**underdrain**  Flow collection and backwash water distribution system used to support the filter bed in most granular media filters. Also called "filter bottom."

**underflow**  The concentrated solids removed from the bottom of a tank or basin.

**underground storage tank (UST)**  A tank used to contain accumulations of regulated substances with 10% or more of its volume underground.

*Uni-Dose*™  Electronic metering pump by Liquid Metronics, Inc.

*Uniflap*  Cast urethane flap valve by Ashbrook Corp.

*Uniflow*  Sloped-bottom settling tanks equipped with chain and flight sludge collector by FMC Corp., MHS Division.

**uniformity coefficient**   Method of characterizing filter sand where the uniformity coefficient is equal to the sieve size in millimeters that will pass 60% of the sand divided by that size passing 10%.

*UniMix*   Vertical or horizontal flocculation unit by Walker Process Equipment Co.

*Uni-Pac*   Water treatment equipment products by Cochrane Environmental Systems.

*UniPro™*   Reverse osmosis pump system by Union Pump Company.

*Unipure*   Heavy metals package waste removal system by Unocal Corp.

*Uni-Scour*   Dual-media filter with self-contained back-wash storage by Smith & Loveless, Inc.

*Unisweep™*   Filter surface wash system by Roberts Filter.

*Unisystem®*   Activated sludge wastewater treatment system by Aeration Industries, Inc.

*Unitank®*   Continuous-flow wastewater treatment system by Seghers-Dinamec, Inc.

*Unitech*   Evaporator and crystallizer product line by Graver Co.

*Unitherm™*   Thermal volatile organic compound oxidation unit by Regenerative Environmental Equipment Co.

*Unitube*   Suction-type sludge removal system for circular clarifiers by Envirex, Inc.

*Uni-Vayor*   Sludge removal conveyor by the former Bowser-Briggs Filtration Co.

*Univer*   Chemicals used to determine water hardness by Hach Company.

*UniversaLevel*   Continuous liquid level transmitter by Drexelbrook Engineering Co.

*Universal RAI*   Rotary positive blowers by Dresser Industries/Roots Division.

**Universal Underdrain®**   Sand filter underdrain by F.B. Leopold Co., Inc.

**Unizone**   Ozone generators by Emery-Trailigaz Ozone Co.

**unleaded gasoline**   Gasoline containing very little or no tetraethyl lead antiknock additives.

**unstable**   Condition describing an element or compound that reacts spontaneously to form other elements or compounds.

**Upcore™**   Upflow countercurrent regeneration process and resins by Dow Chemical.

**UPE®**   Universal Process Equipment Co.

**upflow clarifier**   Clarifier in which flocculated water flows upward through a sludge blanket to obtain floc removal by contact with flocculated solids in the blanket.

**upflow filter**   Granular media filter characterized by an upward flow of liquid through the filter bed.

**upset**   An unexpected disturbance of a process or operation.

**uptake**   The absorption or ingestion of an element or compound in an organism or substance.

**urea**   A soluble nitrogen compound that is a component of mammalian urine.

**USDA**   U.S. Department of Agriculture.

**USEPA**   See "Environmental Protection Agency."

**USGS**   U.S. Geological Survey.

**USP**   U.S. Pharmacopoeia.

**USPHS**   U.S. Public Health Service.

**USP-purified water**   Water meeting U.S. Pharmacopoeia quality requirements that has been purified by distillation, ion exchange, or other suitable process from water complying with U.S. EPA drinking water regulations, and contains no added substances.

**UST**   See "underground storage tank."

**UV**   See "ultraviolet light."

*UV4000*™   Ultraviolet water disinfection system by Trojan Technologies, Inc.

# V

*Vacflush®*   Aeration cleaning system by Jet Tech, Inc.

*Vacuator*   Vacuum-operated system for removal of floating solids and scum by Dorr-Oliver, Inc.

*Vacu-Treat*   Rotary vacuum filter by Envirex, Inc.

**vacuum breaker**   A device that allows air into a water distribution system to prevent backsiphonage.

**vacuum deaerator**   Device operating under vacuum to remove dissolved gases from a liquid.

**vacuum filter**   A sludge dewatering device utilizing a cloth-covered cylindrical drum slowly rotating in a tank of sludge while subject to an internal vacuum. Sludge is drawn to the filter cloth, while water passes through the sludge cake.

**vacuum truck**   A tank truck used to remove wastewater or other liquid wastes in which the air inside the tank is pumped out and the liquid to be removed rushes in to fill the empty space.

**vadose zone**   Designation of the layer of the ground below the surface but above the water table.

**valve**   A device used to regulate flow of fluids through a piping system.

*Valve PAC™*   Valve positioning controller by F.B. Leopold Co., Inc.

**van der Waals' force**   An attractive force apparent in colloidal particles.

*Vanguard®*   Air stripper by Delta Cooling Towers, Inc.

**vapor**   The gaseous phase of a material that is in the solid or liquid state at standard temperature and pressure.

*Vapor Combustor*™ Vapor extraction unit by QED Environmental Systems.

**vapor compression evaporation (VC)** Evaporative system where vapor boiled off in the evaporator is mechanically compressed and reused as the heating medium.

*Vapor Guard* Fabric membrane odor control tank cover by ILC Dover, Inc.

**vapor incinerator** An enclosed combustion device that is used for destroying organic compounds.

**vaporization** The process where a substance changes from a liquid or solid to the gaseous state.

*VaporMate*™ Volatile organic carbon treatment system by North East Environmental Products, Inc.

*Vapor Pacs* Replaceable carbon canister systems for volatile organic compound emission control by Calgon Carbon Corp.

**vapor pressure** The pressure at which equilibrium is established between the liquid and gas phases of a substance.

**vapor recovery system** A system to gather all vapors and gases discharged from a storage tank and process them to prevent their emission to the atmosphere.

*VaPure*® Vapor compression still by Mueller International Sales Corp.

**variable declining-rate filtration** Filter operation where the rate of flow through the filter declines and the level of liquid above the filter rises throughout the filter run.

*Variair* Diffuser by EnviroQuip Inc.

*Vari-Ator* Variable energy controller for floating aerator by Aerators, Inc.

*Vari-Cant*™ Jet aeration system by Jet Tech, Inc.

*Variflo®*   Wastewater distributor nozzle for high-rate trickling filter by Wheelabrator Engineered Systems, Inc., Microfloc Products.

*VariSieve™*   Solids classifier by Krebs Engineers.

*Varivoid™*   High-rate granular media filter by Graver Co.

**vault**   An above- or belowground reinforced concrete structure for storing radioactive or other hazardous waste material.

*V-Auto™*   In-line strainer by Andritz-Ruthner, Inc. (Western Hemisphere) and Andritz Sprout-Bauer S.A. (Eastern Hemisphere).

**VC**   See "vapor compression evaporation."

*VDS*   Vertical drum screen by Jones and Attwood, Inc.

**vector**   An insect or other carrier capable of transmitting a pathogen from one organism to another.

*Vee-wire®*   Wedge-shaped wire by Wheelabrator Engineered Systems, Inc., Johnson Screens.

*Vekton*   Nylon sludge collector sprockets by Norton Performance Plastics.

*Vektor™*   Mixer by Lightnin.

**velocity cap**   The horizontal cap on a vertical, offshore water intake that results in a horizontal inflow, thus reducing fish entrainment.

**velocity gradient (G value)**   A measurement of the degree of mixing imparted to water or wastewater during flocculation.

**velocity head**   The kinetic energy in a hydraulic system.

*VentOXAL®*   Oxygenator system by Liquid Air.

*Vent-Scrub™*   Powdered activated carbon adsorption unit by Wheelabrator Clean Water Systems, Inc.

*VentSorb*   Disposable granular activated carbon filters by Calgon Carbon Corp.

**venturi effect**   An increase in the velocity of a fluid as it passes through a constriction in a pipe or channel.

**venturi meter**   A meter used to measure flows in closed conduits by registering the difference in velocity heads between the entrance and outlet of a contracted throat.

**venturi scrubber**   A wet scrubber used to remove particulates from gaseous emissions.

**vermiculture**   Sludge stabilization and conversion process using earthworms to consume organic matter in sludge.

*VerTech*™   Wet oxidation process by Air Products and Chemicals, Inc.

*Verti-Flo*   Vertical-flow, rectangular tank clarifier by Envirex, Inc.

*Verti-Jet*™   Vertical-pressure leaf filter by U.S. Filter Corp.

*Vertimatic*   Upflow sand filter by U.S. Filter Corp.

*Vertimill*™   Lime slaker by Svedala Industries, Inc.

*Verti-Press*   Inclined screw-type solids/screenings press by Dontech, Inc.

*Verti-Press*   High-pressure filter press by Filtra-Systems.

*Vertiscreen*   Vertical stationary screen by Black Clawson, Shartles Division.

**VFD**   Variable frequency drive.

*Vibra-Matic*™   Vibrating pressure leaf filter cleaning system by U.S. Filter Corp.

*Vibrasieve*®   Vibrating fine screen by Andritz-Ruthner, Inc.

**vibrating screen**   A mechanical screening device whose screening surface is shaken or vibrated to enhance solid/liquid separation. A variation of this screen is used as a solids classifier.

*Viggers Valve*   Automatic sludge blow-off valve by Walker Process Equipment Co.

*Vincent Press*   Horizontal dewatering screw press by Vincent Corp.

*Vinyl Core*   Polyvinyl chloride biological filter media formerly offered by B.F. Goodrich Co.

**VIRALT**   A mathematical model used to calculate the concentration of viruses at the water table and a well after water has been transported through a subsurface media.

*Virchem®*   Corrosion control product by Technical Products Corp.

**virus**   Smallest biological structure capable of reproduction that infect their host, producing disease.

**viscera**   The internal organs of an animal.

*Viscomatic*   Lime slaker by Infilco Degremont, Inc.

**viscosity**   The property of a fluid that offers resistance to flow due to the existence of internal friction.

*Viser*   Sludge management software by OVRC Environmental, Inc.

*Vistex*   Grit removal system by Vulcan Industries, Inc.

*Vitox®*   Oxygen injection system for wastewater treatment plants by BOC Gases.

**vitrified clay**   A kiln-fired clay product used to make pipe and brick.

**VLA**   Very large array. See "array."

**V-notch weir**   A weir with a triangular, V-shaped notch.

**VOC**   See "volatile organic compound."

*VOCarb*   Activated carbon by Wheelabrator Clean Water Systems, Inc.

*VOC Wagon™*   Self-contained thermal oxidizer by NAO Inc.

**volatile**   A substance that evaporates or vaporizes at a relatively low temperature.

**volatile organic compound (VOC)**   Highly evaporative organic compound often found in paints, solvents, and similar products.

**volatile suspended solids (VSS)**   Organic content of suspended solids in a water or wastewater. Determined after heating a sample to 600°C.

*Volcano*™   Continuous downflow sand filter by Lighthouse Separation Systems, Inc.

*Volclay*®   Sodium bentonite soil sealant liner systems by Colloid Environmental Technologies Co.

*Voluette*   Ampules of standard reagent solutions for analytical laboratory use by Hach Company.

*Volumeter*™   Flow monitoring instrument by Marsh-McBirney, Inc.

**volumetric feeder**   Dry chemical feeder that supplies a constant preset or proportional chemical delivery by volume and will not recognize a change in material density.

*VO Nozzle*   Variable-orifice spray nozzle by FMC Corp., MHS Division.

*Vortair*®   Low speed surface aerator by Infilco Degremont, Inc.

**vortex**   A flow with a whirling, rotary motion forming a cavity in the center, toward which particles are drawn.

*Vortex*™   Circular tank grit removal system with turbine-type rotor by Infilco Degremont, Inc.

**vortex grit removal**   Grit removal system relying on a mechanically induced vortex to capture grit solids in the center hopper of a circular tank.

*Vorti-Mix*®   Submerged turbine aerator by Infilco Degremont, Inc.

*Vostrip*™   Air stripping tower by EnviroSystems Supply, Inc.

*VPS*™   Vapor phase odor control system by NuTech Environmental Corp.

*V*Sep*®   Membrane filtration process using vibration to prevent fouling by New Logic International.

**VSR**   Volatile solids reduction.

**VSS**   See "volatile suspended solids."

*VTC*™  Vertical tube coalescing oil/water separator by AFL Industries, Inc.

**VTE**  Vertical tube evaporator.

*VTSH*®  Vertical turbine solids handling pump by Fairbanks Morse Pump Corp.

*VTX*  Vertical vortex pump by Yeomans Chicago Corp.

# W

**WAC**   Weak-acid cation exchanger.

*WAC®*   Polyaluminum chloride flocculant by Water Treatment Solutions Division of Elf Atochem North America.

*WACX*   Treatment equipment product line by Dean Wacx.

**wadi**   Stream or river bed that carries flash flood water.

*Wagner*   Sand filter underdrains by Infilco Degremont, Inc.

**walking beam flocculator**   A mechanical flocculation device whose mixing paddles are attached to a horizontal beam that oscillates to produce a reciprocating motion.

*Wand Inductor*   Contaminated soil remediation system by Biological Systems USA, Inc.

**WAO**   See "wet air oxidation."

**warp**   The vertical wire in woven wire mesh.

**WAS**   See "waste activated sludge."

*Wash Press*   Screenings washer/press by Lakeside Equipment Corp.

*Washpactor*   Sewage screenings washing and compacting unit by Jones and Attwood, Inc.

**waste activated sludge (WAS)**   Excess activated sludge that is discharged from an activated sludge treatment process.

**waste heat evaporator**   An evaporator that uses the heat of a gas turbine, diesel engine jacket water, or exhaust gas.

*Waste Management, Inc.*   Former name of WMX Technologies, Inc.

**waste minimization**    Pollution control plan that includes both pollution source reduction and environmentally sound recycling.

**waste oil emulsion**    A thick, viscous waste that results from a water-in-oil emulsion.

**waste oils**    Lubricating oils that have completed their intended use cycle and must be treated and reused, or receive proper disposal.

**waste stabilization pond**    A pond receiving raw or partially treated wastewater in which stabilization occurs.

*Wastewarrior*    Ultrafiltration wastewater treatment system by Hyde Products, Inc.

**wastewater**    Liquid or waterborne wastes polluted or fouled from households, commercial, or industrial operations, along with any surface water, stormwater, or groundwater infiltration.

*Water Blaze*    Wastewater evaporator by Landa, Inc.

**waterbox**    The chamber at the inlet end of a condenser tubesheet.

*Water Boy®*    Package water treatment plant by Wheelabrator Engineered Systems, Inc., Microfloc Products.

*Water Buffalo™*    Reverse osmosis unit by Mechanical Equipment Co.

*Water Champ*    Chemical induction unit by Gardiner Equipment Co., Inc.

*Water-Chex®*    Chlorine indicator by PyMaH Corp.

*Water Eater*    Wastewater evaporator by Equipment Manufacturing Corp.

**water-effects ratio (WER)**    A test procedure that determines the capacity of site water to mitigate metal toxicity by comparing the toxic endpoint of site water to the toxic endpoint of laboratory water.

**Water Factory 21**   An Orange County, California, treatment plant designed to produce high-quality water from
municipal wastewater for injection into aquifers, creating a coastal barrier and preventing seawater intrusion
of underground water supplies.

**waterflood**   A method of secondary oil recovery in which
water is injected into an oil reservoir to force additional
oil out of the formation into a producing well.

**water-for-injection (WFI)**   Water purified to U.S. Pharmacopoeia standards by distillation, ion exchange, or
other suitable process, containing no additives, and
intended for use as a solvent for preparation of parenteral
solutions.

*Water Free*™   Grit and screenings gate by Tetra Technologies, Inc.

**water hammer**   Condition that may cause damage or rupture to a piping system resulting from a rapid increase
in pressure that occurs in a closed piping system when
the liquid velocity is suddenly changed.

**water hyacinth**   See "hyacinth."

**watermaker**   A common term for a packaged, vapor compression thermal desalination unit.

*Water Maze*®   Wastewater clarifier by Landa, Inc.

**water meter**   A device installed in a pipe that measures
and registers the quantity of water passing through it.

**water reclamation**   The restoration of wastewater to a
state that will allow its beneficial reuse.

**watershed**   The land area that drains into a stream.

**water table**   The upper limit of the portion of the ground
saturated with water.

*Water Tech*   Former equipment manufacturer acquired by
Roberts Filter Manufacturing Co.

**watertube boiler**   Boiler in which the flame and hot combustion gases flow across the outside of the tubes and water is circulated within the tubes.

**water vapor**   The gaseous form of water.

*Waterweb*®   Mesh used in air scrubbers by Misonix, Inc.

**waterworks**   The system of reservoirs, treatment, and pumping by which a water supply is obtained and distributed to a community.

*Wave Oxidation*®   Fluctuating aerobic and anaerobic biological wastewater treatment system by Parkson Corp.

**WBA**   Weak-base anion exchanger.

**weak acid**   A poorly ionized acid that dissociates very little and produces few hydrogen ions in aqueous solution.

*Wedge-Flow*™   Wedgewire filtration products by LEEM Filtration Products, Inc.

*WedgePress*   Belt filter press by the former Gray Engineering Co.

*Wedgewater Filter Bed*™   Gravity sludge dewatering filter bed by Gravity Flow Systems, Inc.

*Wedgewater Sieve*   Static bar screen by Gravity Flow Systems, Inc.

**wedgewire**   General term to describe trapezoidal or V-shaped wire.

*Wedging Roll*   Belt filter press by OVRC Environmental.

**WEF**   Water Environment Federation.

**weft**   The horizontal wire in woven wire mesh. Also called the "shute" wire.

**weighted composite sample**   A composite sample where the sample amounts are based on flow rates at the time of collection; i.e., the higher the flow, the larger the portion of sample taken at that time.

**weir**  A baffle over which water flows.

**weir loading**  The rate of flow out of a basin stated in terms of the volume of liquid passing over a stated length of weir per unit of time.

**weir overflow rate**  A measurement of the volume of water flowing over each unit length of weir per day.

*Weiss*  In-vessel composting system by Taulman Company.

**well**  A bored, drilled, or driven shaft or hole the depth of which is greater than the largest surface dimension.

*Welles Product*s  Former equipment manufacturer acquired by Aerators Inc.

**well screen**  A slotted or perforated well casing that allows passage of water while preventing solids from entering the well.

**well water**  Water from a hole bored, drilled, or otherwise constructed in the ground to tap an aquifer.

*Well Wizard*®  Groundwater monitor and sampling device by QED Environmental Systems.

*Wemco*  Product line of Envirotech Company.

**WER**  See "water-effects ratio."

*WESCO*  Former name of screening product division of Brackett Green, Ltd.

*Westates Carbon*  Carbon products and services by Wheelabrator Clean Air Systems, Inc.

*Westchar*™  Activated carbon by Western Filter Co.

*Westchlor*®  Polyaluminum coagulants by Westwood Chemical Corp.

*Westfalia*  Sludge dewatering decanter centrifuge by Centrico, Inc.

*WestRO*™  Reverse osmosis and membrane products by Western Filter Co.

**WET**  See "whole effluent toxicity."

**wet air oxidation (WAO)**   Process in which sludge and compressed air are pumped into a pressurized reactor and heated to oxidize the volatile solids without vaporizing the liquid.

**wet bulb temperature**   The temperature reading taken from a thermometer with a wetted wick surrounding its bulb.

*Wetec*™   Wet aeration installation system by Aeration Technologies, Inc.

**wetlands**   Surface areas, including swamps, marshes, and bogs, that are inundated or saturated by groundwater frequently enough to support a prevalence of vegetation adapted for life in saturated soil conditions.

**wetlands treatment**   A wastewater treatment system using the aquatic root system of cattails, reeds, and similar plants to treat wastewater applied either above or below the soil surface.

**wet scrubber**   An air pollution control device used to remove particulates and fumes from air by entraining the pollutants in a water spray.

**wet steam**   Steam containing water droplets.

**wetted perimeter**   The length of wetted contact area between a stream of flowing water and the channel that contains it.

**wet well**   A chamber in which water or wastewater is collected and to which the suction of a pump is connected.

**WFI**   See "water-for-injection."

*Wheeler*   Sand filter underdrain by Roberts Filter Manufacturing Co.

*Whirl-Flo*®   Solids handling vortex pump by ITT Marlow.

*Whispair*   Rotary blower and gas pump product line by Dresser Industries/Roots Division.

**white water**   Filtrate from a paper or board forming machine, usually recycled for density control.

*WhiteWater*™   Low-profile air strippers by QED Treatment Systems.

**WHO**   World Health Organization.

**whole effluent toxicity (WET)**   The aggregate toxic effect of an effluent measured directly by a toxicity test.

*Wiese-Flo*®   Self-cleaning filter screen by Wheelabrator Engineered Systems, Wiesemann Products.

*Wiesemann*™   Wiesemann products division of Wheelabrator Engineered Systems.

*William Green*   Former screening equipment manufacturer acquired by Brackett Green, Ltd.

*Willowtech*®   Sludge mixer/blender by Ashbrook Corp.

**windage emission**   Emission of vapors resulting from wind blowing through a free-vented tank and educting some of the saturated vapors.

**windbreak**   A row of trees planted on level land as protection against winds and erosion.

*Windjammer*™   Brush aerator by United Industries, Inc.

**windrow composting**   Sludge composting method where sludge is arranged in long triangular-shaped piles, called windrows, that are mechanically turned and remixed periodically.

*Windrow-Mate*   Compost odor control system by NuTech Environmental Corp.

*Winklepress*®   Belt filter press by Ashbrook Corp. (U.S.) and Bellmer (Europe).

**Winkler titration**   A standard titration method of determining the dissolved oxygen level of a water or wastewater.

**working load**   An allowable recommended tensile load for chains used on conveyors, screens, or other applications of low relative speeds.

**worm**   A shank having at least one complete thread around the pitch surface.

**worm gear**   A gear with teeth cut on an angle to be driven by a worm; used to connect nonparallel, nonintersecting shafts.

**WPCF**   Water Pollution Control Federation. Former name of "Water Environment Federation (WEF)."

**WPCP**   Water pollution control plant.

**WQA**   Water Quality Act of 1987.

*Wring-Dry*   Internally fed rotary fine screen by Schlueter Co.

**WSTA**   Water Science and Technology Association.

**WTP**   Water treatment plant.

*WTS*   Wastewater Treatment Systems, Inc.

**WWEMA**   Water and Wastewater Equipment Manufacturers Association.

**WWTP**   Wastewater treatment plant.

*Wyss®*   Fine-bubble diffuser by Parkson Corp.

# X

*Xentra*  Continuous emissions analyzer by Servomex Co.

**xeric**  Requiring very little moisture to sustain life.

*X-Flo*™  Crossflow-type plastic trickling filter media by American Surfpac Corp.

**XP**  See "explosion proof."

*X-Pruf*™  Explosion proof portable submersible pump by Crane Pumps & Systems.

**X-ray fluorescence**  A technique used in the analysis of water-formed deposits and heavy metals or corrosion products in water.

*Xtractor*™  Sludge removal process by Davis Water & Waste Industries, Inc.

# Y

**yard waste**   Grass clippings, leaves, and miscellaneous vegetative matter.

# Z

**Z Chlor™**  Chlorine and sulfite measurement and control unit by Fischer & Porter.

**ZD**  See "zero discharge."

**zebra mussel**  Freshwater mollusk that can foul water intake screens and piping by attaching itself to a solid structure, eventually restricting flow.

**Zenobox**  Membrane filtration system by Zenon Environmental, Inc.

**Zenofloc**  Polymer control system for belt filter press by Zenon Environmental, Inc.

**Zeo-Karb®**  Sulfonated coal cation exchange process by U.S. Filter Corp.

**Zeol™**  Rotor concentrator for volatile organic compound abatement by Zeol Division-Munters Corp.

**zeolite**  Minerals or synthetic resins that have ion exchange capabilities.

**zeolite softening**  Water softening process using a zeolite resin bed to accomplish ion exchange.

**Zeo-Rex®**  Oxidizing filter for iron and manganese removal by U.S. Filter Corp.

**Zephyr**  Induced-air flotation system by Aeromix Systems, Inc.

**zero discharge (ZD)**  A facility that discharges no material to the environment.

**Zerofuel**  Fluidized bed sludge incinerator by Seghers Dinamec Inc.

**zero liquid discharge (ZLD)**  A facility that discharges no liquid effluent to the environment.

**zero ODP**  Zero ozone depletion potential.

**zero-order reaction**   A reaction in which the rate of change is independent of the concentration of the reactant.

*Zeta-Pak*   Cartridge filter by Alsop Engineering Co.

**zeta potential**   The voltage differential between the surface of the diffuse layer surrounding a colloidal particle and the bulk liquid beyond.

**ZID**   See "zone of initial dilution."

*Zimmerman Process*   Wet air oxidation process by Zimpro Environmental, Inc.

*Zimpress*   Plate and frame sludge press by Zimpro Environmental, Inc.

*Zimpro Passavant Environmental Systems*   Former name of Zimpro Environmental, Inc.

*Zimpro Process*   Thermal sludge conditioning process by Zimpro Environmental, Inc.

**ZLD**   See "zero liquid discharge."

*Z-metal*   Pearlitic malleable iron chain material by Envirex, Inc.

**zone of incorporation**   The depth to which soil on a landfarm is plowed or tilled to receive wastes.

**zone of initial dilution (ZID)**   The part of a lake or river where a discharge from an outfall first mixes with the receiving waters.

**zone of saturation**   The portion of the earth's crust below the water table where the pores are filled with water at greater than atmospheric pressure.

**zooglea**   A gelatinous matrix developed by growing bacteria associated with trickling filter beds and activated sludge floc.

**zooplankton**   Small aquatic animals that possess little or no means of propulsion.

**ZPG**   Zero population growth.

# Appendix: Manufacturers Directory

Abanaki Corp.
P.O. Box 149
Chagrin Falls, OH 44022
Phone: 216-543-7400
Fax: 216-543-7404

ABB Air Preheater, Inc.
P.O. Box 372
Wellsville, NY 14895
Phone: 716-593-2700
Fax: 716-593-2721

ABB Raymond
650 Warrenville Rd.
Lisle, IL 60532
Phone: 708-971-2500
Fax: 708-971-1076

A.B. Marketech, Inc.
9708 North Range Line
  Rd.
Mequon, WI 53092
Phone: 414-255-7448
Fax: 414-241-3443

A-C Compressor Corp.
401 E. South Island St.
Appleton, WI 54915
Phone: 414-738-3080
Fax: 414-738-5167

Acrison, Inc.
20 Empire Blvd.
Moonachie, NJ 07074
Phone: 201-440-8300
Fax: 201-440-4939

ACS Industries, Inc.
14208 Industry Rd.
Houston, TX 77053
Phone: 713-434-0934
Fax: 713-433-6201

ADI Systems, Inc.
1133 Regent St., Ste. 300
Fredericton, NB, Canada,
  E3B 3Z3
Phone: 506-452-7307
Fax: 506-452-7308

Advanced Membrane
  Technology
1305 Calvary Church Rd.
Gainesville, GA 30507
Phone: 770-535-7222
Fax: 770-535-7213

Advanced Microbial
  Systems, Inc.
P.O. Box 239
Shakopee, MN 55379
Phone: 612-445-4251
Fax: 612-445-7233

Advanced Polymer Systems
3696 Haven Ave.
Redwood City, CA 94063
Phone: 415-366-2626
Fax: 415-365-6490

Advanced Sensor Devices,
  Inc.
430 Ferguson Dr.
Mountian View, CA 94043
Phone: 415-960-3007
Fax: 415-960-0127

Advanced Separation
  Technologies
5315 Great Oak Dr.
Lakeland, FL 33801-3180
Phone: 813-687-4460
Fax: 813-687-9362

Advanced Structures, Inc.
2181 Meyers Ave.
Escondido, CA 92029
Phone: 619-738-3000
Fax: 619-738-3059

Advanced Wastewater
  Treatment
Dudley Road, Lye,
  Stourbridge
West Midlands, DY9 8DU
  England
Phone: 0384-892619
Fax: 0384-424081

Adwest Technologies,
  Inc.
803 W. Angus
Orange, CA 92668
Phone: 714-997-8722
Fax: 714-997-8744

Aeration Engineering
  Resources Corp.
76 Webster St.
Worcester, MA 01603-1911
Phone: 508-756-1020
Fax: 508-756-0110

Aeration Industries, Inc.
4100 Peavey Rd.
Chaska, MN 55318
Phone: 612-448-6789
Fax: 612-448-7293

Aeration Research
  Company
3702 Ascot Ln.
Houston, TX 77092
Phone: 713-683-9889
Fax: 713-688-9280

Aeration Technologies,
  Inc.
P.O. Box 488
North Andover, MA
  01845
Phone: 508-475-6385
Fax: 508-475-6387

Aerators, Inc.
11765 Main St.
Roscoe, IL 61073
Phone: 815-623-2111
Fax: 815-623-6416

Aer-O-Flo Environmental, Inc.
1175 Appleby Line, Unit B-2
Burlington, Canada L7L 5H9
Phone: 416-335-8944
Fax: 416-335-8972

Aero-Mod, Inc.
7927 U.S. Hwy 24
Manhattan, KS 66502
Phone: 913-537-4995
Fax: 913-537-0813

Aeromix Systems, Inc.
2611 N. Second St.
Minneapolis, MN 55411-1634
Phone: 612-521-8519
Fax: 612-521-1455

Aeropulse, Inc.
1746 Winchester Rd.
Bensalem, PA 19020
Phone: 215-245-7554
Fax: 215-245-7849

Aero Tec Laboratories, Inc.
Spear Road Industrial Park
Ramsey, NJ 07446-1251
Phone: 201-825-1400
Fax: 201-825-1962

AFL Industries, Inc.
3661 West Blue Heron Blvd.
Riviera Beach, FL 33404
Phone: 561-844-5200
Fax: 561-844-5246

Agar Technologies
P.O. Box 802127
Houston, TX 77043
Phone: 713-464-4451
Fax: 713-464-7741

Ag-Chem Equipment Company
5720 Smetana Dr., Ste. 100
Minnetonka, MN 55343-9688
Phone: 612-933-9006
Fax: 612-933-7432

Agency Environmental, Inc.
2600 Cabover Dr., Ste. J
Hanover, MD 21076
Phone: 410-760-6878
Fax: 410-760-6877

Air-O-Lator Corp.
8100 Paseo
Kansas City, MO 64131
Phone: 816-363-4242
Fax: 816-363-2322

Air Products & Chemicals,
 Inc.
Box 538
Allentown, PA 18195-1501
Phone: 215-481-7380
Fax: 215-481-5900

Airvac, Inc.
P.O. Box 528
Rochester, IN 46975
Phone: 219-223-3980
Fax: 219-223-5566

Alar Engineering Corp.
9651 W. 196th St.
Mokena, IL 60448
Phone: 708-479-6100
Fax: 708-479-9059

Albright & Wilson Americas
P.O. Box 26229
Richmond, VA 23260-6229
Phone: 804-550-4300
Fax: 804-550-4385

Albright & Wilson Ltd.
Oldbury, Warley
W. Midlands B68 0NN,
  England
Phone: 44-21-429-4942
Fax: 44-21-420-5151

Alfa Laval Separation, Inc.
955 Mearns Rd.
Warminster, PA 18974
Phone: 215-443-4000
Fax: 215-443-4234

Allied Colloids, Inc.
P.O. Box 820
Suffolk, VA 23434
Phone: 804-538-3700
Fax: 804-538-0204

Alpine Technology, Inc.
1250 Capitol Texas Hwy,
  2-300
Austin, TX 78792
Phone: 512-329-2809
Fax: 512-328-4792

Alsop Engineering
  Company
P.O. Box 3449
Kingston, NY 12401
Phone: 914-338-0466
Fax: 914-339-1063

Alumina Company Ltd.
Ditton Road Works,
  Widnes
Cheshire, WA8 0PH,
  England
Phone: 051-424-6831
Fax: 051-423-6380

Ambient Technologies,
  Inc.
2999 NE 191 St., Ste. 407
N. Miami Beach, FL 22180
Phone: 305-937-0610
Fax: 305-937-2137

American Bio Tech, Inc.
3223 Harbor Dr.
St. Augustine, FL 32095
Phone: 904-825-1500
Fax: 904-825-1524

American Cyanamid Co.
One Cyanamid Plaza
Wayne, NJ 07470
Phone: 201-831-2000
Fax: 201-839-2761

American International
  Chemical, Inc.
17 Strathmore Rd.
Napic, MA 01760
Phone: 508-665-5805
Fax: 508-665-0927

American Minerals, Inc.
901 E. Eighth Ave., Ste. 200
King of Prussia, PA 19406
Phone: 610-962-5050
Fax: 610-962-5056

American Products
P.O. Box 7455
Port St. Lucie, FL 34985
Phone: 561-340-3866
Fax: 561-340-3831

American Screw Press, Inc.
P.O. Box 600
Oakland, NJ 07436
Phone: 201-337-6382
Fax: 201-337-6187

American Sigma, Inc.
P.O. Box 820
Medina, NY 14103
Phone: 716-798-5580
Fax: 716-798-5599

American Surfpac Corporation
Boot & Chestnut Roads
Downingtown, PA 19335
Phone: 215-873-7600
Fax: 412-826-0399

Ametek Inc., Process
  Division
150 Freeport Rd.
Pittsburgh, PA 15238
Phone: 412-828-9040
Fax: 412-826-0399

Ametek, PMT Division
820 Pennsylvania Blvd.
Feasterville, PA 19053-7886
Phone: 215-355-6900
Fax: 215-355-2937

Amwell, Inc.
1740 Molitor Rd.
Aurora, IL 60505-9990
Phone: 603-898-6900
Fax: 603-898-1647

Anatel Corp.
2200 Central Ave., Ste. F
Boulder, CO 80301
Phone: 303-442-5533
Fax: 303-447-8365

Andritz-Ruthner, Inc.
1010 Commercial Blvd. S.
Arlington, TX 76017
Phone: 817-465-5611
Fax: 817-468-3961

Andritz Sprout-Bauer S.A.
10 Ave. de Concyr, F-45071
Orleans Cedex 2, France
Phone: 33-3851-5738
Fax: 33-3863-1565

A.O. Smith Harvestore
  Products, Inc.
345 Harvestore Dr.
DeKalb, IL 60115
Phone: 815-756-1551
Fax: 815-756-7821

Applied Biochemists, Inc.
6120 W. Douglas Ave.
Milwaukee, WI 53218
Phone: 414-464-8450
Fax: 414-438-5671

Applied Process Equipment
700 Church Rd.
Elgin, IL 60123
Phone: 708-695-9595
Fax: 708-695-9624

Aprotek, Inc.
3316 Corbin Way
Sacramento, CA 95827
Phone: 916-366-6165
Fax: 916-366-7873

APV Crepaco, Inc.
9525 Bryn Mawr Ave.
Rosemount, IL 60018
Phone: 708-678-4300
Fax: 708-678-4407

Aqua-Aerobic Systems, Inc.
P.O. Box 2026
Rockford, IL 61130-0026
Phone: 815-654-2501
Fax: 815-654-2508

Aqua Ben Corporation
1390 N. Manzanita St.
Orange, CA 92667
Phone: 714-771-6040
Fax: 714-771-1465

Aqua-Chem, Inc.
P.O. Box 421
Milwaukee, WI 53201
Phone: 414-359-0600
Fax: 414-577-2723

AquaClear Technologies Corp.
5010 Laguna Park Dr.
Elk Grove, CA 95798
Phone: 916-684-6668
Fax: 916-684-6670

Aqualytics, Inc.
7 Powderhorn Dr.
Warren, NJ 07059
Phone: 908-563-2800
Fax: 908-563-2816

Aqua Magnetics
  International
915-B Harbor Lake Dr.
Safety Harbor, FL 34695
Phone: 813-447-2575
Fax: 813-726-8888

AquaPro, Inc.
258 Shenandoah Dr.
Birmingham, AL 35226
Phone: 205-979-8976
Fax:

Aqua Products, Inc.
25 Rutgers Ave.
Cedar Grove, NJ 07009
Phone: 201-857-2700
Fax: 201-857-8981

Aquarium Systems
8141 Tyler Blvd.
Mentor, OH 44060
Phone: 216-256-1997
Fax: 216-255-8994

Aquarium Systems
43 Rue Gambetta
57400 Sarrebourg, France
Phone: 87-03-6730
Fax: 87-03-1098

Aquatrol Ferr-X Corporation
P.O. Box 531
Toms River, NJ 08754
Phone: 908-505-3100
Fax: 908-505-3038

Argo Scientific
185 Bosstick Blvd.
San Marcos, CA 92069
Phone: 619-727-2620
Fax: 619-727-3380

Arlat, Inc.
150 East Bramalea
Ontario, Canada L6T 1C1
Phone: 416-458-8220
Fax: 416-458-8224

Artisan Industries, Inc.
73 Pond St.
Waltham, MA 02254
Phone: 617-893-6800
Fax: 617-647-0143

Asdor
1255 Nicholson Rd.
Newmarket, Ontario,
  Canada L3Y 7V1
Phone: 905-836-7700
Fax: 905-836-7720

Ashai Glass America, Inc.
450 Lexington Ave., Ste. 1920
New York, NY 10017-3911
Phone: 212-687-4600
Fax: 212-687-4663

Ashbrook Corp.
P.O. Box 16327
Houston, TX 77222
Phone: 713-449-0322
Fax: 713-449-1324

Ashland Chemical, Drew
  Division
One Drew Plaza
Boonton, NJ 07005
Phone: 201-236-7600
Fax: 201-263-4483

Atlantes Chemical Systems, Inc.
303 Silver Spring Rd.
Conroe, TX 77303
Phone: 409-856-4515
Fax: 409-856-4589

Atlantic Ultraviolet Corp.
376 Marcus Blvd.
Hauppauge, NY 11788
Phone: 516-273-0500
Fax: 516-273-0771

Atlas Polar, Hercules Division
P.O. Box 160, Postal Station O
Toronto, Canada M4A 2N3
Phone: 416-751-7740
Fax: 416-751-6475

Auburn International
Eight Electronics Ave.
Danvers, MA 01923
Phone: 508-777-2460
Fax: 508-777-8820

Aurora Pump
800 Airport Rd.
Aurora, IL 60542-1494
Phone: 708-859-7000
Fax: 708-859-7060

Austgen-Biojet Waste Systems
500 Sansome St., Ste. 500
San Francisco, CA 94111-3221
Phone: 415-989-8333
Fax: 415-989-8399

Automatic Filters, Inc.
6363 Wilshire Blvd., Ste. 311
Los Angeles, CA 90048
Phone: 213-651-0530
Fax: 213-651-5236

Automation Products, Inc.
3030 Max Roy St.
Houston, TX 77008-9981
Phone: 713-869-0361
Fax: 713-869-7332

Badger Meter, Inc.
4545 W. Brown Deer Rd.
Milwaukee, WI 53223
Phone: 414-355-0400
Fax: 414-355-7499

Bailey
P.O. Box 8070
Fresno, CA 93747
Phone: 209-252-4491
Fax: 209-453-9030

Baler Equipment Co.
P.O. Box 25150
Portland, OR 97225
Phone: 503-292-4118
Fax: 503-297-5991

Bamag GmbH
P.O. Box 460
Butzbach, D-6308, Germany
Phone: 060-33-839
Fax: 060-33-83-506

BASF
Carl Bosch Strasse 38
Ludwigshafen, D-67056,
  Germany
Phone: 621-60-42258
Fax: 621-60-72944

BCA Industrial Controls,
  Ltd.
9688-187th St.
Surrey, BC, Canada V3T
  4W2
Phone: 604-888-4141
Fax: 604-888-3565

Beaird Industries, Inc.
P.O. Box 31115
Shreveport, LA 71130-1115
Phone: 318-865-6351
Fax: 318-868-1701

Bedminster Bioconversion
  Corp.
12 Exec. Campus 535, Rte.
  38 Ste. 380
Cherry Hill, NJ 08002
Phone: 609-662-2662
Fax: 609-662-4095

BEKO Condensate Systems
  Corp.
1216-D No. Lansing
Tulsa, OK 74106
Phone: 918-585-1223
Fax: 918-585-1224

Bepex Corporation
333 Taft St. NE
Minneapolis, MN 55413
Phone: 612-331-4370
Fax: 612-331-1046

Bethlehem Corp.
P.O. Box 338
Roosevelt, NJ 08555-0338
Phone: 609-443-4545
Fax: 609-259-0644

Betz Laboratories
4636 Somerton Rd.
Trevose, PA 19053
Phone: 215-355-3300
Fax: 215-953-2465

B.F. Goodrich Co.
9911 Brecksville
Brecksville, OH 44141
Phone: 216-447-5000
Fax: 216-447-5250

Bielomatik London, Ltd.
Cotswold St.
London, SE27 0DP England
Phone:
Fax:

Biological Solutions USA,
  Inc.
214 Addicks-Howell Rd.
Houston, TX 77079
Phone: 713-497-1970
Fax: 713-497-1916

Bioprime
P.O. Box 716
Norwich, VT 05055
Phone: 802-649-2227
Fax: 802-649-2238

Bio-Recovery Systems,
  Inc.
2001 Copper Ave.
Las Cruces, NM 88005
Phone: 505-523-0405
Fax: 505-523-1638

Biosab, Inc.
23011 Moulton Pkwy.,
  Ste. G-5
Laguna Hills, CA 92653
Phone: 714-859-6306
Fax: 714-472-9315

Bioscience, Inc.
1550 Valley Ctr. Pkwy.,
  Ste. 140
Bethlehem, PA 18017
Phone: 610-974-9693
Fax: 610-691-2170

Biosystems, Inc.
P.O. Box 158
Rockfall, CT 06481
Phone: 203-344-1079
Fax: 203-344-1068

Biothane Corp.
2500 Broadway, Drawer 5
Camden, NJ 08104
Phone: 609-541-3500
Fax: 609-541-3366

Biothermica International,
  Inc.
P.O. Box 6141
Holyoke, MA 01041
Phone: 514-488-3881
Fax: 514-488-3125

Biotronics Technologies,
  Inc.
W226 N555B Eastmound
  Dr.
Waukesha, WI 53186
Phone: 414-896-2650
Fax: 414-896-2644

Biwater Treatment Ltd.
Biwater Pl., Gregge St.,
  Heywood
Lancashire, England OL10
  2DX
Phone: 0706-367555
Fax: 0706-365598

Black Clawson Co., Shartles
  Division
605 Clark St.
Middletown, OH 45042
Phone: 513-424-7400
Fax: 513-424-1168

BlenTech, Inc.
5460 Elgin St.
Pittsburgh, PA 15206
Phone: 412-661-0569
Fax: 412-661-5997

BOC Gases
575 Mountain Ave.
Murray Hill, NJ 07974
Phone: 908-464-8100
Fax: 908-464-9015

Boliden Intertrade, Inc.
3379 Peachtree, Ste. 300
Atlanta, GA 30326
Phone: 404-239-6700
Fax: 404-239-6701

Brackett Green, Ltd.
Severalls Lane, Colchester
Essex, CO4 4PD England
Phone: 0206-852121
Fax: 0206-844509

Brackett Green USA, Inc.
16850 Saturn Lane, Ste. 140
Houston, TX 77058
Phone: 713-480-7955
Fax: 713-480-8225

Brentwood Industries, Inc.
P.O. Box 605
Reading, PA 19603
Phone: 610-374-5109
Fax: 610-376-6022

Brinecell, Inc.
P.O. Box 27488
Salt Lake City, UT 84127
Phone: 801-973-6400
Fax: 801-973-6463

Brown & Root Environmental
P.O. Box 4574
Houston, TX 77210-4574
Phone: 713-575-3000
Fax: 713-575-4537

Browning-Ferris Industries,
  Inc.
P.O. Box 3151
Houston, TX 77253
Phone: 713-870-8100
Fax: 713-870-7000

Bruner Corp.
P.O. Box 07500
Milwaukee, WI 53207
Phone: 414-747-3700
Fax: 414-747-3812

BTG, Inc.
2364 Park Central Blvd.
Decatur, GA 30035
Phone: 770-981-3998
Fax: 770-987-4126

Budd Company
Franklin Ave. & Grant St.
Phoenixville, PA 19460
Phone: 610-935-0225
Fax: 610-935-7151

Buffalo Technologies, Inc.
P.O. Box 1041
Buffalo, NY 14240
Phone: 716-895-2100
Fax: 716-895-8263

Calciquest, Inc.
1891 I-85 Service Rd.
Charlotte, NC 28208
Phone: 704-394-9868
Fax: 704-394-6784

Calgon Carbon Corporation
P.O. Box 717
Pittsburgh, PA 15230-0717
Phone: 412-787-6700
Fax: 412-787-6713

Calgon Corporation
P.O. Box 1346
Pittsburgh, PA 15230
Phone: 412-777-8000
Fax: 412-777-8154

Callaway Chemical Co.
P.O. Box 2335
Columbus, GA 31902-2335
Phone: 706-576-2000
Fax: 706-596-8756

Capital Controls Co., Inc.
P.O. Box 211
Colmar, PA 18915
Phone: 215-997-4000
Fax: 215-997-4062

Carboline Company
350 Hanley Industrial Ct.
St. Louis, MO 63144
Phone: 314-644-1000
Fax: 314-644-4617

Carbonite Filter Corp.
P.O. Box 1
Delano, PA 18220
Phone: 717-467-3350
Fax: 717-467-7272

Carus Chemical Co., Inc.
P.O. Box 1500
Ottawa, IL 61350
Phone: 815-433-9070
Fax: 815-433-9075

Catalytic Combustion, Inc.
709 21st Ave.
Bloomer, WI 54724
Phone: 715-568-2882
Fax: 715-568-2884

Cathodic Protection
  Services Co.
7700 San Felipe, Ste. 340
Houston, TX 77063
Phone: 715-784-7378
Fax: 715-784-9046

CBI Walker, Inc.
1245 Corporate Blvd., Ste 102
Aurora, IL 60504
Phone: 708-851-7500
Fax: 708-851-9392

Celite Corp.
P.O. Box 519
Lompoc, CA 93438-0519
Phone: 805-735-7791
Fax: 805-735-5699

Centrico, Inc.
100 Fairway Ct.
Northvale, NJ 07647
Phone: 201-767-3900
Fax: 201-767-3416

Centrisys Corp.
1931 Industrial Dr.
Libertyville, IL 60048-0441
Phone: 312-816-9210
Fax: 312-367-1787

C&H Waste Processing
9 Garrett Rd., Yeovil
Somerset, BA20 2TJ England
Phone: 0935-26927
Fax:

Chemineer, Inc.
P.O. Box 1123
Dayton, OH 45401
Phone: 513-454-3200
Fax: 513-454-3379

Chemtrac Systems, Inc.
P.O. Box 921188
Norcross, GA 30092
Phone: 404-449-6233
Fax: 404-447-0889

Cherrington Corp.
Hwy 19 West
Fairfax, MN 55332
Phone: 612-331-3362
Fax: 612-331-3180

Chief Industries, Inc.
Box 2078
Grand Isle, NE 68802-2078
Phone: 308-389-7296
Fax: 308-381-8475

Chlorinators, Inc.
4125 S.W. Martin Hwy, Ste. 2
Palm City, FL 34990
Phone: 561-288-4854
Fax: 561-287-3238

Church & Dwight Co., Inc.
P.O. Box 5297
Princeton, NJ 08543-5297
Phone: 609-683-5900
Fax: 609-497-7176

Cimco Lewis Ozone Systems, Inc.
395 West 1100 North
North Salt Lake, UT 84054
Phone: 800-258-8400
Fax: 801-292-9908

Clack Corp.
P.O. Box 500
Windsor, WI 53598-0500
Phone: 608-846-3010
Fax: 608-846-2586

Clarkson Controls &
  Equipment
12059 Woodbine
Detroit, MI 48239
Phone: 313-255-9110
Fax: 313-255-0406

Claude Laval Corp.
1911 N. Helm
Fresno, CA 93727
Phone: 209-255-1601
Fax: 209-255-8093

Cleaver-Brooks
P.O. Box 421
Milwaukee, WI 53201
Phone: 414-359-0600
Fax: 414-577-3185

CMI-Schneible Co.
P.O. Box 100
Holly, MI 48442
Phone: 810-634-8211
Fax: 810-634-2240

CMS Group, Inc.
140 Snow Blvd., #3, Concord
Ontario, Canada L4K 4C1
Phone: 905-660-7580
Fax: 905-660-0243

Cochrane Environmental
  Systems
Box 60191
King of Prussia, PA
  19406
Phone: 610-265-5050
Fax: 610-265-5432

Colloid Environmental
  Technologies Co.
1500 West Shure Dr.
Arlington Heights, IL
  60004-1434
Phone: 708-392-5800
Fax: 708-506-6150

Composite Technology,
  Inc.
1005 Bluemound Rd.
Fort Worth, TX 76131
Phone: 817-232-1127
Fax: 817-232-1582

Compost Systems Co.
9403 Kenwood Rd.
Cincinnati, OH 45242
Phone: 513-984-0625
Fax: 513-984-4238

Conservatek Industries,
  Inc.
P.O. Box 1678
Conroe, TX 77305
Phone: 409-539-1747
Fax: 409-539-5355

Contra-Shear Engineering, Ltd.
CPO Box 1611
Auckland, New Zealand
Phone: 09-818-6108
Fax: 09-818-6599

Conversion Systems, Inc.
200 Welsh Rd.
Horsham, PA 19044
Phone: 215-784-0990
Fax: 215-784-0971

Cook Screen Technologies,
 Inc.
1292 Glendale-Milford Rd.
Cincinnati, OH 45215
Phone: 513-771-9192
Fax: 513-771-2665

Copa Group
Crest Indus Estate, Tonebridge
Kent, TN12 9QJ England
Phone: 0622-832444
Fax: 0622-831466

Costar
One Alewife Center
Cambridge, MA 02140
Phone: 617-868-6200
Fax: 617-868-2076

Coster Engineering
P.O. Box 3407
Mankato, MN 56002
Phone: 507-625-6621
Fax: 507-625-5883

Crane Pumps & Systems
P.O. Box 603
Piqua, OH 45356
Phone: 513-773-2442
Fax: 513-773-2238

Crisafulli Pump Co.
P.O. Box 1051
Glendive, MT 59330-9985
Phone: 406-365-3393
Fax: 406-365-8088

Cromaglass Corporation
P.O. Box 3215
Williamsport, PA 17701
Phone: 717-326-3396
Fax: 717-326-6426

Crompton & Knowles Corp.
P.O. Box 33157
Charlotte, NC 28233
Phone: 704-372-5890
Fax: 704-332-8785

Cross Machine, Inc.
167 Glen Ave.
Berlin, NH 03570
Phone: 603-752-6111
Fax: 603-752-3825

CSF Treatment Systems,
 Inc.
P.O. Box 19390
Portland, OR 97280
Phone: 503-644-8220
Fax: 503-526-0775

CSM Environmental Systems, Inc.
Brooklyn Navy Yard, Bldg. 12
Brooklyn, NY 11205
Phone: 718-522-7000
Fax: 718-852-1686

Culligan International Corp.
One Culligan Pkwy.
Northbrook, IL 60062
Phone: 708-205-6000
Fax: 708-205-6005

Cuno Separations Systems Division
50 Kerry Pl.
Norwood, MA 02062
Phone: 617-769-6112
Fax: 617-769-3274

Cytec Industries, Inc.
Five Garret Mountain Plaza
West Paterson, NJ 07424
Phone: 201-357-3100
Fax: 201-357-3066

Dacar Chemical Company
1007 McCartney St.
Pittsburgh, PA 15220
Phone: 412-921-3620
Fax: 412-921-4478

D&A Instrument Co.
218 Polk St., No. 298
Port Townsend, WA 98368
Phone: 206-385-0272
Fax: 206-385-0460

DAS International, Inc.
673 Exton Commons
Exton, PA 19341
Phone: 215-524-2420
Fax: 215-524-5407

Davis Industries, Process Division
P.O. Box 29
Tallevast, FL 34270-0029
Phone: 813-355-2971
Fax: 813-351-4756

Davis Water & Waste Industries, Inc.
P.O. Box 1419
Thomasville, GA 31792
Phone: 912-226-5733
Fax: 912-228-0312

DBS Manufacturing, Inc.
5421 Hillside Dr.
Forest Park, GA 30050
Phone: 404-768-2131
Fax: 404-761-6360

Dean Wacx
26 Water Ln., Wimslow
Cheshire, SK9 5AA, England
Phone: 0625-529895
Fax: 0625-539425

Delta Chemical Corp.
2601 Cannery Ave.
Baltimore, MD 21226
Phone: 410-354-0100
Fax: 410-354-1021

Delta Cooling Towers, Inc.
134 Clinton Rd.
Fairfield, NJ 07004-2970
Phone: 201-227-0300
Fax: 201-227-0458

Denver Equipment Company
P.O. Box 340
Colorado Springs, CO 80901
Phone: 719-471-3443
Fax: 719-471-4469

Derrick Corp.
590 Duke Rd.
Buffalo, NY 14225
Phone: 716-683-9010
Fax: 716-683-4991

Desalination Systems, Inc.
760 Shadowridge Dr.
Vista, CA 92083-7986
Phone: 619-598-3334
Fax: 619-598-3335

Dexsil Corp.
One Hamden Park Dr.
Hamden, CT 06517
Phone: 203-288-3509
Fax: 203-248-6523

DHV Water B.V.
P.O. Box 484, 3800 AL
Amersfoort, the Netherlands
Phone: 31-33682200
Fax: 31-33682301

Diagenex, Inc.
288 Lindbergh Ave.
Livermore, CA 94550
Phone: 510-606-5600
Fax: 510-606-5643

Diemme USA
1866 Colonial Village Ln.,
  Ste. 111
Lancaster, PA 17601
Phone: 717-394-1977
Fax: 717-394-1973

Dieterich Standard
P.O. Box 9000
Boulder, CO 80301
Phone: 303-530-9600
Fax: 303-530-7064

Diversified Remediation
  Controls
21801 Industrial Blvd.
Rogers, MN 55374
Phone: 612-438-3000
Fax: 612-428-3660

Dontec, Inc.
76 Center Dr.
Gilberts, IL 60136
Phone: 708-428-8222
Fax: 708-428-6855

Dorr-Oliver, Inc.
P.O. Box 3819
Milford, CT 06460-8719
Phone: 203-876-5400
Fax: 203-876-5432

Douglas Engineering
1015 Shary Cir.
Concord, CA 94518
Phone: 510-827-4100
Fax: 510-827-4999

Dow Chemical Co.
P.O. Box 1206
Midland, MI 48641-9940
Phone: 517-636-1000
Fax: 517-638-9783

Dresser Industries/Roots Div.
900 West Mount St.
Connersville, IN 47331
Phone: 317-827-9200
Fax: 317-825-7669

Drexelbrook Engineering
  Co.
205 Keith Valley Rd.
Horsham, PA 19044
Phone: 215-674-1234
Fax: 215-674-2731

D.R. Sperry & Company
112 North Grant St.
North Aurora, IL 60542
Phone: 708-892-4361
Fax: 708-892-1664

Dunlop, Ltd.
Moody Ln.
Grimsby, S. Humberside,
  DN31 2SP. England
Phone: 0472-359281
Fax: 0472-362948

Duriron Company, Filter
  Systems
9542 Hardpan Rd.
Angola, NY 14006
Phone: 716-549-2500
Fax: 716-549-3950

Dürr Environmental Division
40600 Plymouth Rd.
Plymouth, MI 48170
Phone: 313-459-6800
Fax: 313-459-5837

Dustex Corp.
P.O. Box 7368
Charlotte, NC 28241-7368
Phone: 704-588-2030
Fax: 704-588-2032

Dynaphore, Inc.
2709 Willard Rd.
Richmond, VA 23294
Phone: 804-288-7109
Fax: 804-282-1325

Eaglebrook, Inc.
1150 Junction Ave.
Schererville, IN 46375
Phone: 219-322-2560
Fax: 219-322-8533

Earth Science Laboratories, Inc.
201 Holiday Blvd., Ste. 402
Covington, LA 70433
Phone: 504-893-6100
Fax: 504-893-3993

E. Beaudrey & Co.
14 Blvd. Orano
75018 Paris, France
Phone: 42-57-14-35
Fax: 42-64-74-62

Eco Equipment
3300 Blvd. de Enterp, Terrebonne
Quebec, Canada J6X 4J8
Phone: 514-477-7880
Fax: 514-477-7879

Ecolochem, Inc.
P.O. Box 12775
Norfolk, VA 23502
Phone: 804-855-9000
Fax: 804-855-1478

Ecolotech Corp.
P.O. Box 2470
LaGrange, IL 60525
Phone: 708-352-2277
Fax: 708-579-9842

Eco Purification Systems, Inc.
8813 Waltham Woods Rd., Ste. 304
Baltimore, MD 21234
Phone: 410-882-1566
Fax: 410-882-2910

EET Inc.
4710 Bellaire Blvd., Ste. 300
Houston, TX 77401
Phone: 713-662-0727
Fax: 713-662-2322

EG&G Biofiltration
North St.
Saugerties, NY 12477
Phone: 914-246-3711
Fax: 914-246-3802

EG&G Environmental
681 Anderson Dr., Ste. 1400
Pittsburgh, PA 15220
Phone: 412-920-5401
Fax: 412-920-5402

EG&G Rotron, Inc.
North St.
Saugerties, NY 12477
Phone: 914-246-3401
Fax: 914-246-3802

Eichrom Industries, Inc.
8205 South Cass, Ste. 107
Darien, IL 60561
Phone: 708-963-0320
Fax: 708-963-0381

E. I. du Pont de Nemours, Inc.
P.O. Box 6101
Newark, DE 19714-6101
Phone: 302-999-4600
Fax: 302-451-9686

Eimco Process Equipment Co.
P.O. Box 300
Salt Lake City, UT 84110
Phone: 801-526-2000
Fax: 801-526-2014

Ejector Systems, Inc.
910 National Ave.
Addison, IL 60101-9812
Phone: 708-543-2214
Fax: 708-543-2014

Ekofinn Bioclere
1403 South 348th St.
Federal Way, WA 98003
Phone: 206-661-6128
Fax: 206-874-5247

ELCAT Corp.
14299 Wicks Blvd.
San Leandro, CA 94577
Phone: 415-895-6663
Fax: 415-638-6667

Electrocatalytic, Ltd.
Severnbridge Industrial Estate
Gwent, England NP6 4YN
Phone: 0291-423833
Fax: 0291-423836

Elf Atochem North America,
  Inc.
2000 Market St.
Philadelphia, PA 19103
Phone: 215-419-7000
Fax: 215-419-5230

Eltech International Corp.
1110 Industrial Blvd.
Sugar Land, TX 77478
Phone: 713-240-6770
Fax: 713-240-6762

Emery-Trailigaz Ozone Co.
11501 Goldcoast Dr.
Cincinnati, OH 45249
Phone: 513-530-7702
Fax: 513-530-7711

EMI, Inc.
P.O. Box 912
Clinton, CT 06413
Phone: 860-669-1199
Fax: 860-669-7461

EM Industries, Inc.
480 Democrat Rd.
Gibbstown, NJ 08027
Phone: 609-354-9200
Fax: 609-423-4389

EMO France
B.P. 22 - Z.Z. Planche Fagline
35740 Pace, France
Phone: 99-60-63-63
Fax: 99-60-20-87

Endress+Hauser
2350 Endress Pl.
Greenwood, IN 46143
Phone: 317-535-7138
Fax: 317-535-8498

Engineering Specialties, Inc.
P.O. Box 2960
Covington, LA 70434
Phone: 504-892-0071
Fax: 504-892-0474

Entoleter, Inc.
P.O. Box 1919
New Haven, CT 06509
Phone: 203-787-3575
Fax: 203-787-1492

Envirex, Inc.
P.O. Box 1604
Waukesha, WI 53187
Phone: 414-541-0141
Fax: 414-541-0120

Enviro-Care Company
5614 West Grand Ave.
Chicago, IL 60639
Phone: 312-745-7773
Fax: 312-745-8383

Enviroflow, Inc.
12181 Balls Ford Rd.
Manassas, VA 22110
Phone: 703-368-9067
Fax: 703-368-7336

EnviroMetrics
6360 Raton Cir.
Beaumont, TX 77708
Phone: 409-892-0097
Fax: 409-892-6762

Environetics, Inc.
1201 Commerce St.
Lockport, IL 60441
Phone: 815-838-8331
Fax: 815-838-8336

Environmental Construction Ltd.
Danebury Ct., Old Sarum Park
Salisbury, SP4 6EB England
Phone:
Fax:

Environmental Dynamics, Inc.
4509 I-70 Dr., S.E.
Columbia, MO 65201-6709
Phone: 314-474-9456
Fax: 314-474-6988

Environmental Engineering Ltd.
Little London, Spalding
Lincolnshire, RE11 2UE England
Phone: 0775-768964
Fax: 0775-710294

Enviropax, Inc.
P.O. Box 65039
Salt Lake City, UT 84165
Phone: 801-263-8880
Fax: 801-263-8898

Enviroquip, Inc.
P.O. Box 9069
Austin, TX 78766
Phone: 512-218-3200
Fax: 512-218-3277

EnviroQuip International, Inc.
8506 Beechmont Ave.
Cincinnati, OH 45255
Phone: 513-388-4100
Fax: 513-388-4111

EnviroSystems Supply, Inc.
3806 N. 29th Ave.
Hollywood, FL 33020
Phone: 954-925-9993
Fax: 954-925-9996

Envirotech Company
P.O. Box 209
Salt Lake City, UT 84110-0209
Phone: 801-359-8731
Fax: 801-355-9303

Envirotech Systems Corp.
999 West Hastings St.
Vancouver, BC, Canada
  V6C 2W2
Phone: 604-685-8496
Fax: 604-683-2256

EPG Companies, Inc.
Box 224
Maple Grove, MN 55369
Phone: 612-424-2613
Fax: 612-493-4812

Equipment Manufacturing
  Corp.
2615 Pacific Park Dr.
Whittier, CA 90601
Phone: 310-908-7696
Fax: 310-908-7698

ERM Group
855 Springdale Dr.
Exton, PA 19341
Phone: 610-524-3500
Fax: 610-524-7335

Euroquip Fabrication, Ltd.
Telford Way, Kettering
Nothants, NN16 8YZ England
Phone: 0536-81629
Fax: 0536-410193

Everfilt Corp.
3167 Progress Cir.
Mira Loma, CA 91752
Phone: 909-360-8380
Fax: 909-360-8384

Fairbanks Morse Pump Corp.
3601 Fairbanks Ave.
Kansas City, KS 66109-0999
Phone: 913-371-5000
Fax: 913-371-2272

Farmer Automatic of America
P.O. Box 39
Register, GA 30452
Phone: 912-681-2763
Fax: 912-681-1096

Farr Co.
2221 Park Pl.
El Segundo, CA 90245
Phone: 310-536-6300
Fax: 310-643-9086

F.B. Leopold Co., Inc.
227 S. Division St.
Zelienople, PA 16063
Phone: 412-452-6300
Fax: 412-452-1377

FE3, Inc.
P.O. Box 808
Celina, TX 75009
Phone: 214-382-2381
Fax: 214-382-3211

Ferro Corp.
P.O. Box 389
East Rochester, NY 14445
Phone: 716-586-8770
Fax: 716-586-7154

Fibercor
14605 28th Ave. North
Minneapolis, MN 55447
Phone: 612-553-3300
Fax: 612-553-3387

Filtra-Systems
P.O. Box 1007
Wixom, MI 48393-1007
Phone: 313-669-0300
Fax: 313-669-0308

Filtronics, Inc.
1157 N. Grove St.
Anaheim, CA 92806
Phone: 714-630-5040
Fax: 714-630-1160

Fischer & Porter
125 E. County Line Rd.
Warminster, PA 18974
Phone: 215-674-6000
Fax: 215-674-7183

Fisher Controls
  International, Inc.
P.O. Box 190
Marshalltown, IA 50158
Phone: 515-754-3000
Fax: 515-754-2830

Flo-Trend Systems, Inc.
707 Lehman
Houston, TX 77018-1513
Phone: 713-699-0152
Fax: 713-699-8054

Flottweg
7095 Industrial Rd.
Florence, KY 41042-6270
Phone: 606-283-0200
Fax: 606-283-9678

Floway Pumps, Inc.
2494 S. Railroad Ave.
Fresno, CA 93706
Phone: 209-442-4000
Fax: 209-442-3098

Flow Process Technologies, Inc.
11917 Windfern Rd.
Houston, TX 77064
Phone: 713-469-2777
Fax: 713-469-2232

Fluid Dynamics, Inc.
6595 ODell Pl., Ste. E
Boulder, CO 80301-3316
Phone: 303-530-7300
Fax: 303-530-7754

Fluid Systems Corp.
10054 Old Grove Rd.
San Diego, CA 92131
Phone: 619-695-3840
Fax: 619-695-2176

FMC Corp., MHS Division
400 Highpoint Dr.
Chalfont, PA 18914
Phone: 215-822-4300
Fax: 215-822-4342

FMC Corp., Peroxygen
  Chemicals
1735 Market St.
Philadelphia, PA 19103
Phone: 215-299-6000
Fax: 215-299-5819

FMC Corp., Process
  Additives Division
Tenax Rd., Trafford
  Park
Manchester, M17 1WT
  England
Phone: 061-872-2323
Fax: 061-875-3175

FMC Corp., Process
  Additives Division
1735 Market St.
Philadelphia, PA 19103
Phone: 215-299-6000
Fax: 215-299-6121

Force Flow Equipment
3467 Golden Gate Way
Lafayette, CA 94549
Phone: 510-284-2200
Fax: 510-284-5945

Ford Hall Co., Inc.
P.O. Box 54312
Lexington, KY 40555
Phone: 606-624-3320
Fax: 606-624-3320

Formulabs, Inc.
P.O. Box 1869
Piqua, OH 45356
Phone: 513-773-0600
Fax: 513-773-7994

Framco Environmental
  Technologies
1856 Apex Rd.
Sarasota, FL 34240
Phone: 941-377-3600
Fax: 941-377-6900

Franklin Miller
60 Okner Pkwy.
Livingston, NJ 07039
Phone: 201-535-9200
Fax: 201-535-6269

Futura Coatings, Inc.
9200 Latty Ave.
Hazelwood, MO 63042-2805
Phone: 314-521-4100
Fax: 314-521-7255

GA Industries, Inc.
9025 Marshall Rd.
Mars, PA 16046
Phone: 412-776-1020
Fax: 412-776-1254

Gardiner Equipment
  Company, Inc.
6911 Breen Rd., B-1
Houston, TX 77086
Phone: 713-999-5193
Fax: 713-999-5197

Gardner-Denver
1800 Gardner Expwy.
Quincy, IL 62301
Phone: 217-222-5400
Fax: 217-224-7814

Garland Manufacturing Co.
P.O. Box 538
Saco, ME 04072
Phone: 207-283-3693
Fax: 207-283-3693

Gauld Equipment Sales Co.
P.O. Box 1129
Theodore, AL 36582
Phone: 334-653-8558
Fax: 334-653-0533

General Chemical Corporation
90 E. Halsey Rd.
Parsippany, NJ 07054
Phone: 201-515-0900
Fax: 201-515-4461

General Filter Co.
600 Arrasmith Trail
Ames, IA 50010
Phone: 515-232-4121
Fax: 515-232-2571

GeoPure Continental
2300 NW 71st Place
Gainesville, FL 32653
Phone: 904-376-7833
Fax: 904-376-7660

Geosource Ltd.
4100 One Shell Square
New Orleans, LA 70139
Phone: 504-566-7722
Fax: 504-566-7729

G-H Systems, Inc.
P.O. Box 1089
LaPorte, TX 77571
Phone: 713-471-5689
Fax: 713-471-5835

G.E.T. Industries, Inc.
P.O. Box 640
Brampton, Ontario, Canada
 L6V 2L6
Phone: 416-451-9900
Fax: 416-451-5376

Girard Industries
6531 N. Eldridge
 Pkwy.
Houston, TX 77041
Phone: 713-466-3100
Fax: 713-466-8050

Gladwall Engineering,
 Ltd.
6213 Wagner Rd.
Edmonton, Alberta,
 Canada T6E 4N4
Phone: 403-465-5451
Fax: 403-465-9929

Glegg Water Conditioning
 Co.
29 Royal Rd.
Guelph, Ontario, Canada
 N1H 1G2
Phone: 519-836-0500
Fax: 519-837-3067

Goal Line Environmental
 Technologies
P.O. Box 58324
Los Angeles, CA 90058
Phone: 213-233-2224
Fax: 213-233-7428

Goodwin Pumps of America
One Floodgate Rd.
Bridgeport, NJ 08014
Phone: 609-467-3636
Fax: 609-467-4841

Grace TEC Systems
830 Prosper Rd.
DePere, WI 54115-0030
Phone: 414-336-5715
Fax: 414-336-3404

Graver Company
2720 U.S. Hwy 22
Union, NJ 07083
Phone: 908-964-2400
Fax: 908-964-7770

Gravity Flow Systems, Inc.
P.O. Box 525
Carbondale, PA 18407
Phone: 717-282-6036
Fax: 717-282-3081

Greasby
500 Technology Ct.
Smyrna, GA 30082
Phone: 770-319-9999
Fax: 770-319-0336

Great Lakes Instruments, Inc.
9020 West Dean Rd.
Milwaukee, WI 53224
Phone: 414-355-3601
Fax: 414-355-8346

Great Lakes International,
Inc.
1905 Kearny Ave.
Racine, WI 53403
Phone: 414-634-2386
Fax: 414-634-6259

Gundle Lining Systems, Inc.
19103 Gundle Rd.
Houston, TX 77073
Phone: 713-443-8564
Fax: 713-875-6010

Guzzler Manufacturing, Inc.
575 N. 37th St.
Birmingham, AL 35222
Phone: 205-591-2477
Fax: 205-591-2495

Hach Company
P.O. Box 389
Loveland, CO 80539-0389
Phone: 997-669-2932
Fax: 907-669-2932

Hadley Industries
5900 West Fourth St.
Ludington, MI 49341
Phone: 616-845-0537
Fax: 616-843-3882

Hankin Environmental
  Services
P.O. Box 935
Somerville, NJ 08876
Phone: 908-722-9595
Fax: 908-722-9514

Harmsco Industrial
  Filters
P.O. Box 14066
N. Palm Beach, FL
  33408
Phone: 561-848-9628
Fax: 561-845-2474

Harris Waste Management
  Group
200 Clover Reach Dr.
Peachtree City, GA 30269
Phone: 404-631-7290
Fax: 404-631-7299

Haynes International,
  Inc.
1020 W. Park Ave.
Kokomo, IN 46904
Phone: 317-456-6000
Fax: 317-456-6905

Hayward Industrial
  Products, Inc.
900 Fairmount Ave.
Elizabeth, NJ 07207-9990
Phone: 908-351-5400
Fax: 908-351-7893

Hazleton Environmental
  Products
P.O. Box 488
Hazleton, PA 18201-0488
Phone: 717-454-7515
Fax: 717-454-7520

Healy-Ruff Company
2485 North Fairview Ave.
St. Paul, MN 55113
Phone: 612-633-7522
Fax: 612-633-2671

Heinkel Filtering Systems,
  Inc.
P.O. Box 503
Bridgeport, NJ 08014
Phone: 609-467-3399
Fax: 609-467-1010

Hellmut Geiger GmbH &
  Co. KG
P.O. Box 210163
76151 Karlsruhe, Germany
Phone: 49-721-5001-0
Fax: 49-721-5001-213

Hercules Systems Ltd.
2 Shieling Ct., N. Folds
  Rd.
Corby, Northants, NN18
  9QD England
Phone: 0536-460210
Fax: 0536-460232

Heyl & Patterson, Inc.
P.O. Box 36
Pittsburgh, PA 15230-0036
Phone: 412-788-6900
Fax: 412-788-6913

HF Scientific, Inc.
3170 Metro Pkwy.
Ft. Myers, FL 33916-7537
Phone: 941-337-2116
Fax: 941-332-7643

H&H Pump and Dredge Co.
520 Hwy 322
Clarksdale, MS 38614
Phone: 601-627-9631
Fax: 601-627-9660

H.I.L. Technology, Inc.
P.O. Box 366
Scarborough, ME 04074-0366
Phone: 207-883-9100
Fax: 207-883-1010

Hinde Engineering Co.
P.O. Box 737
Aromas, CA 95004
Phone: 408-726-2644
Fax: 408-726-2609

Hindon Corporation
2055 Bee's Ferry Rd.
Charleston, NC 29414
Phone: 803-763-6616
Fax: 803-763-2338

Hinsilblon, Ltd.
1025 T. Jefferson NW,
#411W
Washington, D.C. 20007
Phone: 202-625-0777
Fax: 202-625-0888

Hitachi Maxco, Ltd.
1630 Albritton Dr.
Kennesaw, GA 30144
Phone: 404-424-9350
Fax: 404-424-9145

Hitachi Metals America,
Ltd.
2400 Westchester Ave.
Purchase, NY 10577
Phone: 914-694-9200
Fax: 914-694-9279

Homa Pumps, Inc.
18 Elmcroft
Stamford, CT 06902
Phone: 203-327-6365
Fax: 203-356-1064

Houseman Limited
The Priory, Burnham
Buckinghamshire SL1 7LS,
England
Phone: 44-1628-604488
Fax: 44-1628-666381

H2O Waste-Tec Operations
Horsfield Way, Bredbury
Park
Stockport, SK6 2SU
England
Phone: 061-406-7111
Fax: 061-406-7222

Howe-Baker Engineers, Inc.
P.O. Box 956
Tyler, TX 75710
Phone: 903-597-0311
Fax: 903-597-8670

Hoyt Corp., Westport
Environ
251 Forge Rd.
Westport, MA 06477
Phone: 508-636-8811
Fax: 508-636-2088

Hubert Stavoren B.V.
Kooyweg 20, 8715 EP
Stavoren, the Netherlands
Phone: 31-5149-1625
Fax: 31-5149-2198

Humboldt Decanter, Inc.
3883 Steve Reynolds Blvd.
Norcross, GA 30093
Phone: 404-564-7300
Fax: 404-416-9377

Hungerford & Terry, Inc.
P.O. Box 650
Clayton, NJ 08312
Phone: 609-881-3200
Fax: 609-881-6859

Hychem, Inc.
10014 North Dale Maybry
  Hwy.
Tampa, FL 33618
Phone: 813-963-6214
Fax: 813-960-0175

Hycor Corporation
29850 North Hwy 41
Lake Bluff, IL 60044
Phone: 708-473-3700
Fax: 708-473-0477

Hyde Products, Inc.
28045 Ranney Pkwy.
Westlake, OH 44145
Phone: 216-871-4885
Fax: 216-871-1143

Hydranautics
8444 Mirilani Dr.
San Diego, CA 92126
Phone: 619-536-2500
Fax: 619-536-2578

Hydro-Aerobics, Inc.
1615 State Route 131
Milford, OH 45150
Phone: 513-575-2800
Fax: 513-575-2896

HydroCal Company
22732 Granite Way, Ste. A
Laguna Hills, CA 92653
Phone: 714-455-0765
Fax: 714-455-0764

Hydroflo Corp.
6100 Easton
Plumsteadville, PA 18949
Phone: 215-766-7867
Fax: 215-776-8290

Hydro Gate Corp.
6101 North Dexter St.
Commerce City, CO 80022
Phone: 303-288-7873
Fax: 303-287-8531

Hydro Group, Inc., Ranney
  Division
2 North State St.
Westerville, OH 43081
Phone: 614-882-3104
Fax: 614-882-3071

Hydrolab Corp.
P.O. Box 50116
Austin, TX 78763
Phone: 512-255-8841
Fax: 512-255-3106

Hydropress Wallander &
  Co., AB
Heljesvagen 4, Box 125
S-437 22 Lindome, Sweden
Phone: 4631-995050
Fax: 4631-995133

IBERO Anlagentechnik
GmbH
Heinich-KrummStrabe 7
D-6050 Offenbach-am-
Main, Germany
Phone:
Fax:

ICI Katalco
2 Transam Plaza Dr.,
Ste., 230
Oak Brook Terrace, IL
60181
Phone: 708-268-6300
Fax: 708-268-9797

IDEXX Laboratories, Inc.
One IDEXX Dr.
Westbrook, ME 04092
Phone: 207-856-0300
Fax: 207-856-0346

Idreco USA, Ltd.
3494 Progress Rd., Ste. B
Bensalem, PA 19020
Phone: 215-638-2111
Fax: 215-638-2114

ILC Dover, Inc.
P.O. Box 266
Frederica, DE 19946
Phone: 302-335-3911
Fax: 302-335-0762

Industrial Environmental
Supply
P.O. Box 36210
Greensboro, NC 27416-
6210
Phone: 910-274-4817
Fax: 910-274-9499

Industrial Filter & Pump
Mfg. Co.
5900 Ogden Ave.
Cicero, IL 60650-3888
Phone: 708-656-7800
Fax: 708-656-7816

Infilco Degremont, Inc.
P.O. Box 71390
Richmond, VA 23255-1390
Phone: 804-756-7600
Fax: 804-756-7645

Ingersoll-Dresser Pump
Co.
P.O. Box 91
Taneytown, MD 21787
Phone: 410-756-3278
Fax: 410-756-2275

Innova-Tech, Inc.
4 Lee Blvd.
Malvern, PA 19355
Phone: 610-640-1310
Fax: 610-640-2670

In-Situ, Inc.
210 S. Third St.
Laramie, WY 82070
Phone: 307-742-8213
Fax: 307-721-7598

Integrated Environmental
  Solutions
3787 Old Middleburg Rd.,
  Ste. 3
Jacksonville, FL 33210
Phone: 904-778-1188
Fax: 904-778-0201

International Dioxide,
  Inc.
136 Central Ave.
Clark, NJ 07066
Phone: 908-499-9660
Fax: 908-388-3648

International Filter Media
P.O. Box 216
Hazelton, PA 18201-0458
Phone: 717-459-1491
Fax: 717-455-7510

International Pigging
  Products
P.O. Box 692005
Houston, TX 77269
Phone: 713-351-6688
Fax: 713-255-2385

International Shredder, Inc.
P.O. Box 526
Cohoes, NY 12047
Phone: 518-785-0064
Fax: 518-785-1001

IN USA, Inc.
100 Crescent Rd. Unit 1B
Needham, MA 02194
Phone: 617-444-2929
Fax: 617-444-9229

Invincible AirFlow Systems
P.O. Box 380
Baltic, OH 43804
Phone: 216-897-3200
Fax: 216-897-3400

Ionics, Inc.
65 Grove St.
Watertown, MA 02172
Phone: 617-926-2500
Fax: 617-926-4304

Isco, Inc.
P.O. Box 82531
Linclon, NE 68501-2531
Phone: 402-474-2233
Fax: 402-474-6685

ITT Flygt Corp.
P.O. Box 1004
Trumbull, CT 06611-0943
Phone: 203-380-4700
Fax: 203-380-4705

ITT Marlow
1150 Tennessee Ave.
Cincinnati, OH 45229
Phone: 513-482-2500
Fax: 513-482-2569

IX Services Co.
1102 Holly St.
Las Cruces, NM 88005
Phone: 505-526-2838
Fax: 505-526-2838

Jaeger Products, Inc.
1611 Peach Leaf
Houston, TX 77039
Phone: 713-449-9500
Fax: 713-449-9400

Javex Manufacturing Corp.
255 Wicksteed Ave.
Toronto, Ontario, Canada
  M4H 1G8
Phone: 416-421-6000
Fax: 416-425-9320

Jay R. Smith Mfg. Co.
2781 Gunter Park Dr., East
Montgomery, AL 36109-1405
Phone: 334-272-7396
Fax: 334-272-7396

JBS Instruments
311 D St.
West Sacramento, CA 95605
Phone: 916-372-0534
Fax: 916-372-1624

JDV Equipment Corp.
P.O. Box 471
Montville, NJ 07045
Phone: 201-335-4740
Fax: 201-335-4702

Jeffrey Chain Corp.
2307 Maden Dr.
Morristown, TN 37813
Phone: 615-586-1951
Fax: 615-581-2399

Jeffrey Division/Indresco
P.O. Box 387
Woodruff, SC 29388
Phone: 803-476-7523
Fax: 803-476-7765

Jet, Inc.
750 Alpha Dr.
Cleveland, OH 44143
Phone: 216-461-2000
Fax: 216-442-9008

Jet Tech, Inc.
P.O. Box 13306
Edwardsville, KS 66113
Phone: 913-422-7600
Fax: 913-422-7667

John Meunier, Inc.
6290 Perinault
Montreal, Quebec, Canada
  H4K 1K5
Phone: 514-334-7230
Fax: 514-334-5070

John Zink Company
P.O. Box 21220
Tulsa, OK 74121-1220
Phone: 918-234-1800
Fax: 918-234-1975

Jones and Attwood, Inc.
1931 Industrial Dr.
Libertyville, IL 60048
Phone: 708-367-5480
Fax: 708-367-1787

JWC Environmental
290 Paularino Ave.
Costa Mesa, CA 92626
Phone: 714-833-3888
Fax: 714-833-8858

JWI, Inc.
2155 112th Ave.
Holland, MI 49423
Phone: 616-772-9011
Fax: 616-772-4516

Kason Corporation
1301 E. Linden Ave.
Linden, NJ 07036
Phone: 908-486-8140
Fax: 908-486-8598

KCC Corrosion Control Co.
4010 Trey Rd.
Houston, TX 77084
Phone: 713-550-1199
Fax: 713-550-9097

Kem-Tron
3050 Post Oak Blvd.,
  Ste. 699
Houston, TX 77056
Phone: 713-621-0075
Fax: 713-621-0080

Ketema, Inc.
P.O. Box 1406
El Cajon, CA 92022
Phone: 619-449-0202
Fax: 619-449-0883

Keystone Engineering &
  Treatment
P.O. Box 360
Black Diamond, WA
  98010-0491
Phone: 206-886-1396
Fax: 206-886-2480

Kimre, Inc.
P.O. Box 570846
Perrine, FL 33257-0846
Phone: 305-233-4249
Fax: 305-233-8687

Kinetico Engineered
  Systems, Inc.
10845 Kinsman Rd.
Newbury, OH 44065
Phone: 216-564-5397
Fax: 216-338-8694

King Lee Technologies
8949 Kenamar Dr., Bldg. 107
San Diego, CA 92121-2453
Phone: 619-693-4062
Fax: 619-693-4917

Klenzoid, Inc.
912 Spring Mill Ave.
Conshohocken, PA 19428
Phone: 215-825-9494
Fax: 215-825-0238

Knapp Polly Pig, Inc.
1209 Hardy
Houston, TX 77020
Phone: 713-222-0146
Fax: 713-222-7403

Koch Membrane Systems, Inc.
850 Main St.
Wilmington, MA 01887
Phone: 508-657-4250
Fax: 508-658-2883

Koflo Corp.
309 Cary Point Rd.
Cary, IL 60013
Phone: 708-516-3700
Fax: 708-516-3724

Komax Systems, Inc.
P.O. Box 1323
Wilmington, CA 90748-1323
Phone: 213-830-4320
Fax: 213-830-9826

Komline-Sanderson
  Engineering
12 Holland Ave.
Peapack, NJ 07977-0257
Phone: 908-234-1000
Fax: 908-234-9487

Kopcke Industrie B.V.
Delta Industrieweg 36
32 LX Stellendam, the
  Netherlands
Phone: 01879-2988
Fax: 01879-2781

KRC (Hewitt) Inc.
P.O. Box 68
Neenah, WI 54957
Phone: 414-722-7713
Fax: 414-725-8615

Krebs Engineers
1205 Chrysler Dr.
Menlo Park, CA 94025
Phone: 415-325-0751
Fax: 415-326-7048

KriStar Enterprises
422 Larkfield, Ste. 271
Santa Rosa, CA 95403
Phone: 707-525-0973
Fax: 707-579-8819

Krofta Engineering Corporation
P.O. Box 972
Lenox, MA 01240
Phone: 413-637-0740
Fax: 413-637-0768

Krüger, Inc.
401 Harrison Oaks Blvd.,
Ste. 100
Cary, NC 27513
Phone: 919-677-8310
Fax: 919-677-0082

K-Tron North America
P.O. Box 888
Pitman, NJ 08071
Phone: 609-589-0500
Fax: 609-589-8113

Kvaerner Eureka USA
P.O. Box 129
Bel Air, MD 21014
Phone: 410-838-9042
Fax: 410-838-9254

Lake Aid Systems
510 E. Main St.
Bismarck, ND 58501
Phone: 701-222-8331
Fax: 701-222-2773

Lakeside Equipment Corp.
P.O. Box 8448
Bartlett, IL 60103
Phone: 708-837-5640
Fax: 708-837-5647

Lamson Corporation
P.O. Box 4857
Syracuse, NY 13221
Phone: 315-433-5513
Fax: 315-433-5451

Landa, Inc.
13705 NE Airport Way
Portland, OR 97230-1048
Phone: 503-255-5980
Fax: 503-255-1509

Land Combustion
2525-B Pearl Buck Rd.
Bristol, PA 19007
Phone: 215-781-0810
Fax: 215-781-0798

Landustrie Sneek B.V.
P.O. Box 199
86 AD Sneek, the
Netherlands
Phone: 05150-11411
Fax: 05150-12398

Lantec Products, Inc.
5308 Derry Ave., Unit E
Agoura Hills, CA 91301
Phone: 818-707-2285
Fax: 818-707-9367

Larox Inc.
9730 Patuxent Woods Dr.
Columbia, MD 21046
Phone: 301-381-3314
Fax: 301-381-4490

LCI Corp.
P.O. Box 16348
Charlotte, NC 28297-9984
Phone: 704-394-8341
Fax: 704-393-8590

Leeds & Northrup
Sumneytown Pike
North Wales, PA 19454
Phone: 215-699-2000
Fax: 215-699-3702

LEEM Filtration Products,
  Inc.
124 Christie Ave.
Mahwah, NJ 07430
Phone: 201-529-4747
Fax: 201-529-2131

Lemacon Techniek B.V.
P.O. Box 53
6666 ZH, Heteren, the
  Netherlands
Phone: 8306-23266
Fax: 8306-23455

Lemna Corp.
1408 Northland Dr., Ste.
  310
St. Paul, MN 55120
Phone: 612-688-0836
Fax: 612-688-8813

Licon, Inc.
200 E. Government St.,
  Ste. 130
Pensacola, FL 32501
Phone: 904-434-5088
Fax: 904-438-2040

Lighthouse Separation Systems
Box 840
Dahlonega, GA 30533
Phone: 706-864-8644
Fax: 706-864-8677

Lightnin
P.O. Box 1370
Rochester, NY 14603
Phone: 716-436-5550
Fax: 716-436-5589

Liquid Air
5230 S. East Ave.
Countryside, IL 60525
Phone: 708-482-8400
Fax: 708-579-7702

Liquid Carbonic
810 Jorie Blvd.
Oak Brook, IL 60521-2216
Phone: 708-572-7500
Fax: 708-572-7935

Liquid Metronics, Inc.
8 Post Office Square
Acton, MA 01720
Phone: 508-263-9800
Fax: 508-264-9172

Liquid-Solids Separation Corp.
P.O. Box 9
Northvale, NJ 07647
Phone: 201-784-1570
Fax: 201-784-1575

LIST, Inc.
40 Nagog Park
Acton, MA 01720
Phone: 508-635-9521
Fax: 508-635-0570

Longwood Engineering
  Company
Parkwood Mills,
  Huddersfield
W. Yorkshire, HD3 4TP
  England
Phone: 0484-642011
Fax: 0484-642935

Lowry Aeration Systems,
  Inc.
P.O. Box 14209
Research Triangle Park, NC
  27709
Phone: 919-544-9080
Fax: 919-544-8720

LWT, Inc.
Box 250
Somerset, WI 54025
Phone: 715-247-3322
Fax: 715-247-3934

Magnet Machinery, Inc.
North Maple St.
Florence, MA 01060
Phone: 413-586-4030
Fax: 413-586-6905

Mahr Maschinenbau
  Ges.m.b.H.
Salzgries 1
1010 Vienna, Austria
Phone: 43-1-533-553127
Fax: 43-1-533-553125

Mahr USA
11415 Brittmore Park Dr.
Houston, TX 77041-
  6919
Phone: 713-466-9026
Fax: 713-466-4473

Markland Specialty
  Engineering
48 Shaft Rd., Rexdale
Toronto, Ontario, Canada
  M9W 4M2
Phone: 416-244-4980
Fax: 416-244-2287

Marolf, Inc.
15500 49th St. N.
Clearwater, FL 34622
Phone: 813-536-5991
Fax: 813-531-0915

Marsh-McBirney, Inc.
4539 Metropolitan Ct.
Frederick, MD 21701-
  8364
Phone: 301-874-5599
Fax: 301-874-2172

Martin Marietta Specialties, Inc.
P.O. Box 15470
Baltimore, MD 21220-0470
Phone: 410-780-5500
Fax: 410-780-5777

Mass Transfer, Inc.
5583 Ridge Ave.
Cincinnati, OH 45213
Phone: 513-531-1986
Fax: 513-531-1987

Mass Transfer Systems, Inc.
100 Waldron Rd.
Fall River, MA 02720-4732
Phone: 508-679-6770
Fax: 508-672-5779

Matheson Gas Products
166 Keystone Dr.
Montgomeryville, PA 18936
Phone: 215-641-2700
Fax:

Matrix Desalination, Inc.
3295 SW 11th Ave.
Fort Lauderdale, FL 33315
Phone: 954-524-5120
Fax: 954-524-5216

Matt-Son, Inc.
28W005 Industrial Ave
Barrington, IL 60010
Phone: 708-382-7810
Fax: 708-382-5814

Mazzei Injector Corp.
500 Rooster Dr.
Bakersfield, CA 93307-9555
Phone: 805-363-6500
Fax: 805-363-7500

McLanahan Corp.
200 Wall St.
Hollidaysburg, PA 16648
Phone: 814-695-9807
Fax: 814-695-6684

McTigue Industries, Inc.
Box 928
Mitchell, SD 57301
Phone: 605-996-1162
Fax: 605-996-1908

Mechanical Equipment Company
861 Carondelet St.
New Orleans, LA 70130
Phone: 504-523-7271
Fax: 504-525-4846

Medina Products, Bio Division
P.O. Box 309
Hondo, TX 78861
Phone: 210-426-3011
Fax: 210-426-2288

Megator Corp.
562 Alpha Dr.
Pittsburgh, PA 15238
Phone: 412-963-9200
Fax: 412-963-9214

Membrex, Inc.
155 Route 146 West
Fairfield, NJ 07004
Phone: 201-575-8388
Fax: 201-575-7011

Memtec America Corporation
5 West Aylesbury Rd.
Timonium, MD 21093
Phone: 410-252-0800
Fax: 410-628-0017

Meridian Diagnostics, Inc.
3741 River Hills Dr.
Cincinnati, OH 45244
Phone: 513-271-3700
Fax: 513-271-0124

Mer-Made Filter, Inc.
85 College Ave.
N. Tarrytown, NY 10591
Phone: 914-332-0440
Fax: 914-332-0204

Merrick Industries, Inc.
10 Arthur Dr.
Lynn Haven, FL 32444
Phone: 904-265-3611
Fax: 904-265-9768

Met-Pro Corp.
P.O. Box 1299
West Chester, PA 19380
Phone: 215-692-3500
Fax: 215-692-0197

MGI Pumps, Inc.
847 Industrial Dr.
Bensenville, IL 60106
Phone: 708-595-8711
Fax: 708-595-8832

Microbics Corporation
2232 Rutherford Rd.
Carlsbad, CA 92008
Phone: 619-438-8282
Fax: 619-438-2980

MIE, Inc.
1 Federal St., #2
Billerica, MA 01821-3500
Phone: 508-663-7900
Fax: 508-663-4890

Miles, Inc.
Mobay Rd.
Pittsburgh, PA 15208-9741
Phone: 412-777-2000
Fax: 412-777-2447

Millgard Environmental
  Corp.
P.O. Box 2708
Livonia, MI 48151
Phone: 313-261-9760
Fax: 313-261-7417

Millipore Corp.
80 Ashby Rd.
Bedford, MA 01730
Phone: 617-275-9200
Fax: 617-533-8878

Milltronics, Inc.
709 Stadium Dr. East
Arlington, TX 76011-9870
Phone: 817-277-3543
Fax: 817-277-3894

Milton Roy Company
201 Ivyland Rd.
Ivyland, PA 18974
Phone: 215-441-0800
Fax: 215-441-8620

Misonix, Inc.
1938 New Highway
Farmingdale, NY 11735
Phone: 516-694-9555
Fax: 516-694-9412

MoDo-Chemetics
1818 Cornwall Ave.
Vancouver, BC, Canada
  V6J 1C7
Phone: 604-734-1200
Fax: 604-734-0340

ModuTank, Inc.
41-04 35th Ave.
Long Island City, NY 11101
Phone: 718-392-1112
Fax: 718-786-1008

Monitek Technologies, Inc.
1495 Zephyr Ave.
Hayward, CA 94544
Phone: 510-471-8300
Fax: 510-471-8647

Monsanto Enviro-Chem
  Systems
P.O. Box 14547
St. Louis, MO 63178
Phone: 314-275-5700
Fax: 314-275-5701

Mooers Products, Inc.
5554 N. Navajo Ave.
Milwaukee, WI 53217
Phone: 414-228-6655
Fax: 414-228-9909

Mt. Fury Co., Inc.
1460 19th Ave., NW
Issaquah, WA 98027
Phone: 206-391-0747
Fax: 206-391-9708

Mueller Company
P.O. Box 828
Springfield, MO 65801-0828
Phone: 417-831-3000
Fax: 417-831-3528

Munters
P.O. Box 6428
Fort Myers, FL 33911
Phone: 941-936-1555
Fax: 941-936-6582

M&W Industries, Inc.
P.O. Box 952
Rural Hall, NC 27045
Phone: 910-969-9526
Fax: 910-969-2156

Nalco Chemical Company
One Nalco Center
Naperville, IL 60566-1024
Phone: 708-305-1000
Fax: 708-305-2900

NAO Inc.
1284 Sedgley Ave.
Philadelphia, PA 19134
Phone: 215-743-5300
Fax: 215-743-3018

National Fluid Separators,
  Inc.
827 Hanley Industrial Ct.
St. Louis, MO 63144-1402
Phone: 314-968-2838
Fax: 314-968-4773

National Seal Co.
1245 Corporate Blvd., Ste. 300
Aurora, IL 60504
Phone: 708-898-1161
Fax: 708-898-3461

Nature Plus, Inc.
52 Lakeview Ave.
New Canaan, CT 06840
Phone: 203-972-1100
Fax: 203-966-2200

Neogen Corp.
620 Lesher Pl.
Lansing, MI 48912
Phone: 517-372-9200
Fax: 517-372-2006

Neotronics
P.O. Box 2100
Flowery Branch, GA
  30542-2100
Phone: 770-967-2196
Fax: 770-967-1854

Netzsch, Inc.
119 Pickering Way
Exton, PA 19341-1393
Phone: 610-363-8010
Fax: 610-363-0971

New Logic International
1295 67th St.
Emeryville, CA 94608
Phone: 510-655-7305
Fax: 510-655-7307

Norair Engineering Corp.
337 Brightseat Rd., Ste. 200
Landover, MD 20785
Phone: 301-499-2202
Fax: 301-499-1342

Norchem Industries
760 N. Frontage Rd.
Willowbrook, IL 60521
Phone: 708-654-4900
Fax: 708-654-4905

Norit Americas, Inc.
1050 Crown Pointe Pkwy.,
  #1500
Atlanta, GA 30338
Phone: 770-512-4610
Fax: 770-512-4622

Norkot Mfg. Co.
P.O. Box 89
Bottineau, ND 58318
Phone: 701-228-3757
Fax: 701-228-2127

North East Environmental
  Products
17 Technology Dr.
West Lebanon, NH 03784
Phone: 603-298-7061
Fax: 603-298-7063

Norton Company
P.O. Box 350
Akron, OH 44309-0350
Phone: 216-673-5860
Fax: 216-677-7245

Norton Performance Plastics
150 Dey Rd.
Wayne, NJ 07470
Phone: 201-696-4700
Fax: 201-696-4056

NRG, Inc.
P.O. Box 306
Ardmore, PA 19003-9998
Phone: 610-896-6850
Fax: 610-649-5083

NSW Corp.
530 Gregory Ave.
Roanoke, VA 24016
Phone: 703-981-0362
Fax: 703-345-8421

NuTech Environmental Corp.
5350 N. Washington St.
Denver, CO 80216-1951
Phone: 303-295-3702
Fax: 303-295-6145

N-Viro Energy Systems, Inc.
3450 W. Central Ave., Ste.
  328
Toledo, OH 43606
Phone: 419-535-6374
Fax: 419-535-7008

NWW Acumem, Inc.
2724 Loker Ave., West
Carlsbad, CA 92008
Phone: 619-929-7500
Fax: 619-929-7510

Odor Management, Inc.
2720 Nevada Ave. North
Minneapolis, MN 55427-9707
Phone: 612-546-9730
Fax: 612-546-8539

Ohmicron
375 Pheasant Run
Newtown, PA 18940
Phone: 215-860-5115
Fax: 215-860-5213

Olin Corp., Chemicals Group
120 Long Ridge Rd.
Stamford, CT 06904-1355
Phone: 203-356-2000
Fax: 203-356-3595

Omega Engineering, Inc.
P.O. Box 4047
Stamford, CT 06907-0047
Phone: 203-359-1660
Fax: 203-359-7700

Omnidata International, Inc.
P.O. Box 448
Logan, UT 84323-0448
Phone: 801-753-7760
Fax: 801-753-6756

Onsite-Offsite, Inc.
2040 Central Ave.
Duarte, CA 91010
Phone: 818-303-2229
Fax: 818-303-1489

Oritex Corporation
780 N. Loren Ave.
Azusa, CA 91702
Phone: 818-815-9900
Fax: 818-815-4065

Orival, Inc.
40 N. Van Brunt St.
Englewood, NJ 07631
Phone: 201-568-3311
Fax: 201-568-1916

ORS Environmental Systems
32 Mill St.
Greenville, NH 03048
Phone: 603-878-2500
Fax: 603-878-3866

Osmonics, Inc.
5951 Clearwater Dr.
Minnetonka, MN 55343-8995
Phone: 612-933-2277
Fax: 612-933-0141

Osprey Biotechnics
2530 B Trailmate Dr.
Sarasota, FL 34243
Phone: 813-755-7770
Fax: 813-755-0626

Otto H. York Company, Inc.
P.O. Box 3100
Parsippany, NJ 07054
Phone: 201-299-9200
Fax: 201-299-9401

Outomec USA, Inc.
1014 East 900 South, Ste. 1
Salt Lake City, UT 84105
Phone: 801-364-8334
Fax: 801-355-2653

OVRC Environmental Inc.
100 RiverPoint Corporate
  Ctr. Dr.
Birmingham, AL 35243-3331
Phone: 205-969-3010
Fax: 205-969-3020

Ozone Pure Water, Inc.
5330 Ashton Ct.
Sarasota, FL 34233
Phone: 813-923-8528
Fax: 813-923-8231

Ozone Research & Equipment
4953 West Missouri Ave.
Phoenix, AZ 85301-6100
Phone: 602-931-7332
Fax: 602-931-7727

Ozonia North America
P.O. Box 330
Lodi, NJ 07644
Phone: 201-778-2131
Fax: 201-778-2357

Parkson Corp.
P.O. Box 408399
Fort Lauderdale, FL 33340-
8399
Phone: 954-974-6610
Fax: 954-974-6182

Particle Measuring Systems,
Inc.
1855 South 57th Ct.
Boulder, CO 80301
Phone: 303-443-7100
Fax: 303-449-6870

Passavant-Werke AG & Co.
D-6209
Aabergen 7, Germany
Phone: 49-6120-282434
Fax: 49-6120-282576

Patterson Candy International
21 The Mall, Ealing
London, W5 2PU England
Phone: 01-579-1311
Fax: 01-840-6180

Patterson Pump Co.
P.O. Box 790
Toccoa, GA 30577
Phone: 706-886-2101
Fax: 706-886-0023

Peabody TecTank
P.O. Box 996
Parsons, KS 67357
Phone: 316-421-0200
Fax: 316-421-9122

Pepcon Systems, Inc.
3770 Howard Hughes
Pkwy., #340
Las Vegas, NV 89109
Phone: 702-735-2324
Fax: 702-735-9456

Peroxidation Systems, Inc.
5151 East Broadway, Ste.
600
Tucson, AZ 85711
Phone: 602-790-8383
Fax: 602-790-8008

Phelps Dodge Refining Co.
P.O. Box 20001
El Paso, TX 79998
Phone: 915-775-8826
Fax: 915-775-8350

Philadelphia Mixer Corporation
1221 East Main St.
Palmyra, PA 17078
Phone: 717-838-1341
Fax: 717-838-8146

Phoenix Process
  Equipment Co.
2402 Watterson Trail
Louisville, KY 40229
Phone: 502-499-6198
Fax: 502-499-1079

Photovac Inc.
25-B Jefryn Blvd.
  West
Deer Park, NY 11729
Phone: 516-254-4199
Fax: 516-254-4284

Pica
16, rue Trzel
92300 Levallois, France
Phone: 47-39-60-40
Fax: 47-39-34-92

Pitt-Des Moines, Inc.
3400 Grand Ave.,
  Neville Island
Pittsburgh, PA 15225
Phone: 412-331-3000
Fax: 412-331-3188

Plasti-Fab, Inc.
P.O. Box 100
Tualatin, OR 97062
Phone: 503-692-5460
Fax: 503-692-1145

Polcon Sales Ltd.
3378 Douglas St., #203
Victoria, BC, Canada V8Z
  3L3
Phone: 604-658-4241
Fax: 604-658-4216

Pollution Control, Inc.
7529 Sussex Dr.
Florence, KY 41042-
  2211
Phone: 606-282-2200
Fax: 606-282-2205

Pollution Control
  Engineering, Inc.
6 Autry, Ste. B
Irvine, CA 92718-2708
Phone: 714-830-8383
Fax: 714-830-3738

Polybac Corp.
3894 Courtney St.
Bethlehem, PA 18017
Phone: 610-867-7338
Fax: 610-861-0991

Polymetrics
1210 Elko Dr.
Sunnyvale, CA 94089
Phone: 408-734-9820
Fax: 408-734-3058

Polypure, Inc.
One Gatehall Dr.
Parsippany, NJ 07054
Phone: 201-292-2900
Fax: 201-292-5295

Positive Flow Systems, Inc.
45 Hymus, Pte. Claire
Montreal, Quebec, Canada
 H9R 4T2
Phone: 514-694-6460
Fax: 514-426-2444

PPC Biofilter
3000 E. Marshall
Longview, TX 75601
Phone: 903-758-3395
Fax: 903-758-6487

PPG Industries, Inc.
One PPG Place
Pittsburgh, PA 15272
Phone: 412-434-3131
Fax: 412-434-4578

PQ Corporation
P.O. Box 840
Valley Forge, PA 19482
Phone: 610-651-4200
Fax: 610-251-5249

Praxair Inc.
39 Old Ridgebury Rd.
Danbury, CT 06810-5113
Phone: 203-837-2000
Fax: 203-794-2055

Premier Services Corp.
7521 Engle Rd., Ste. 415
Middleburg Heights, OH
 44130
Phone: 216-234-4600
Fax: 216-234-5772

Presto-Tek Corp.
2909 Tanager Ave.
Los Angeles, CA 90040
Phone: 213-728-9800
Fax: 213-725-3036

Probiotic Solution
201 S. Roosevelt
Chandler, AZ 85226
Phone: 602-961-1220
Fax: 602-961-3501

Process Combustion Corp.
P.O. Box 12866
Pittsburgh, PA 15241
Phone: 412-655-0955
Fax: 412-655-0961

Pro-Ent, Inc.
P.O. Box 23611
Jacksonville, FL 32241
Phone: 904-737-3536
Fax: 904-737-3537

ProGuard Filtration Systems
P.O. Box 678
Nowata, OK 74048
Phone: 918-273-2208
Fax: 918-273-2101

ProMinent Fluid Controls,
  Inc.
136 Industry Dr.
Pittsburgh, PA 15275
Phone: 412-787-2484
Fax: 412-787-0704

Prototech Company
32 Fremont St.
Needham, MA 02194
Phone: 617-444-5188
Fax: 617-444-0130

Pulsafeeder
77 Ridgefield Rd.
Rochester, NY 14523
Phone: 716-292-8000
Fax: 716-424-5619

Pump Engineering, Inc.
1004 W. Hurd Rd.
Monroe, MI 48161
Phone: 313-242-1772
Fax: 313-242-9777

Purac Engineering, Inc.
4550 New Linden Hill Rd.
Wilmington, DE 19808
Phone: 302-996-0545
Fax: 302-996-0544

Purafil, Inc.
P.O. Box 287
Oconomowoc, WI 53066
Phone: 414-567-0094
Fax: 414-567-3177

Purator Waagner-Brio
Postfach 53, A-1234 Wien
Lembockgasse 49, Austria
Phone: 1/816 07-0
Fax: 1/816 07-232

Purestream, Inc.
P.O. Box 68
Florence, KY 41022-0068
Phone: 606-371-9898
Fax: 606-371-3577

Purolite Co.
150 Monument Rd.
Bala Cynwyd, PA 19004
Phone: 215-668-9090
Fax: 215-668-8139.

Purus Corp.
2713 North First St.
San Jose, CA 95134
Phone: 408-955-1000
Fax: 408-955-1010

PWT Americas Corp.
17400 E. Chestnut Ave.
City of Industry, CA 91749
Phone: 818-912-5411
Fax: 818-913-1034

PyMaH Corp.
500 Route 202 North
Flemington, NJ 08822
Phone: 908-788-4000
Fax: 908-788-4101

QED Environmental Systems
P.O. Box 3726
Ann Arbor, MI 48106
Phone: 313-995-2547
Fax: 313-995-1170

Quad Environmental
  Technologies
3605 Woodhead Dr., Ste 103
Northbrook, IL 60062
Phone: 708-564-5070
Fax: 708-564-5606

Quantum Technologies, Inc.
1632 Enterprise Pkwy.
Twinsburg, OH 44087
Phone: 216-425-7880
Fax: 216-425-0955

Ralph B. Carter Co.
P.O. Box 340
Hackensack, NJ 07602
Phone: 201-342-3030
Fax: 201-342-3668

Ray Products
RD 3, Box 383
Troy, PA 16947-0383
Phone: 215-635-2883
Fax: 215-635-2883

RaySolv, Inc.
P.O. Box 207
Bound Brook, NJ 08805
Phone: 908-981-0500
Fax: 908-356-3629

RDP Company
2495 Blvd. of the Generals
Norristown, PA 19403-5236
Phone: 610-650-9900
Fax: 610-650-9070

Recra Environmental
10 Hazelwood, Ste. 106
Amherst, NY 14228
Phone: 716-691-2600
Fax: 716-691-2617

Red Fox Environmental, Inc.
P.O. Box 10539
New Iberia, LA 70562-0539
Phone: 318-367-0952
Fax: 318-235-2499

Red Valve Co., Inc.
700 N. Bell Ave.
Carnegie, PA 15106
Phone: 412-279-0044
Fax: 412-279-7878

Refractron Technologies Corp.
5750 Stuart Ave.
Newark, NJ 14513
Phone: 315-331-6222
Fax: 315-331-7254

Regenerative Environmental
  Equipment
P.O. Box 1500
Somerville, NJ 08876
Phone: 908-685-4600
Fax: 908-685-4181

Reheis, Inc.
235 Snyder Ave.
Berkeley Heights, NJ 07922
Phone: 908-464-1500
Fax: 908-464-7726

Reidel Smith Environmental
Box 5007
Portland, OR 97208
Phone: 503-286-4656
Fax: 503-283-2602

Resources Conservation Co.
3006 Northup Way
Bellevue, WA 98004-1407
Phone: 206-828-2400
Fax: 206-828-0526

Reuter-Stokes
Edison Park
Twinsburg, OH 44087
Phone: 216-963-2477
Fax: 216-425-4045

R.E. Wright Associates, Inc.
3240 Schoolhouse Rd.
Middletown, PA 17057
Phone: 717-944-5501
Fax: 717-944-5642

Rhone-Poulenc Basic
  Chemicals Co.
1 Corporate Dr., Box 881
Shelton, CT 06484
Phone: 203-925-3627
Fax: 203-925-3300

Rio Linda Chemical Co.,
  Inc.
410 N. 10th St.
Sacramento, CA 95814
Phone: 916-443-4939
Fax: 916-443-5145

Robbins & Myers, Inc.
P.O. Box 960
Springfield, OH 45501
Phone: 513-327-3510
Fax: 513-327-3064

Roberts Filter
  Manufacturing Co.
P.O. Box 167
Darby, PA 19023
Phone: 610-583-3131
Fax: 610-583-0117

Rochem Separation
  Systems, Inc.
3904 Del Amo Blvd.,
  Ste. 801
Torrence, CA 90503
Phone: 310-370-3160
Fax: 310-370-4988

Rochester Midland
P.O. Box 1515
Rochester, NY 14603-
  1515
Phone: 716-336-2200
Fax: 716-266-1606

Rodney Hunt Company
46 Mill St.
Orange, MA 01364
Phone: 508-544-2511
Fax: 508-544-7204

Roediger Pittsburgh, Inc.
3812 Rt. 8
Allison Park, PA 15101
Phone: 412-487-6010
Fax: 412-487-6005

Rohm & Haas Co.
Lennig House, 2, Mason's Ave.
Croydon, CR0 3NB England
Phone: 081-686-8844
Fax:

Rohm & Haas Co.
Independence Mall West
Philadelphia, PA 19105
Phone: 215-592-3000
Fax: 215-592-3377

Ropur AG
4142 Munchebstein 1
Switzerland
Phone: 41-61-415-8710
Fax: 41-61-415-8720

Rosemount Analytical, Inc.
2400 Barranca Pkwy.
Irvine, CA 92714
Phone: 714-863-1181
Fax: 714-474-7250

Rosenmund, Inc.
St. James Ct., Warrington
Chesire, WA4 6PS England
Phone: 0925-52621
Fax: 0925-416790

Rosenmund, Inc.
9110 Forsyth Park Dr.
Charlotte, NC 28273
Phone: 704-587-0440
Fax: 704-588-6866

Rosewater Engineering Ltd.
Stonebroom, Alfreton
Derbyshire, DE5 6LQ England
Phone: 0773-875017
Fax:

Roto-Sieve AB
Hjorthagsgatan 10
S-413 17, Goteborg, Sweden
Phone: 031-427890
Fax: 031-422070

R.P. Adams Co., Inc.
P.O. Box 963
Buffalo, NY 14240-0963
Phone: 716-877-2608
Fax: 716-877-9385

R. Spane GmbH
Schafmatt 5
79618 Reinfelden, Germany
Phone: 49-7623-1084
Fax: 49-7623-20660

Rubber Millers, Inc.
709 S. Caton
Baltimore, MD 21229
Phone: 410-947-8400
Fax: 410-233-6537

Rupprecht & Patashnick
  Co., Inc.
25 Corporate Cir.
Albany, NY 12203
Phone: 518-452-0065
Fax: 518-452-0067

SanTech, Inc.
25 Commercial Dr.
Wrentham, MA 02093
Phone: 508-384-3181
Fax: 508-384-5346

Schlicher & Schuell
P.O. Box 2012
Keene, NH 03431
Phone: 603-352-3810
Fax: 603-357-3627

Schloss Engineered Equipment
10555 E. Dartmouth #230
Aurora, CO 80014
Phone: 303-695-4500
Fax: 303-695-4507

Schlueter Co.
P.O. Box 548
Janesville, WI 53547
Phone: 608-755-0740
Fax: 608-755-0332

Schreiber Corp.
100 Schreiber Dr.
Trussville, AL 35173
Phone: 205-655-7466
Fax: 205-655-7669

Schumacher Filters, Ltd.
41 Cavendish St.
Sheffield, S3 7RZ England
Phone: 742-766-654
Fax:

SciCorp Systems, Inc.
19 Churchill Dr.
Barrie, Ontario, Canada
  L4M 6E7
Phone: 705-733-2626
Fax: 705-733-2618

Scienco/FAST Systems
3240 North Broadway
St. Louis, MO 63147-3515
Phone: 314-621-2536
Fax: 314-621-1952

Scoti-Zahner, Inc.
P.O. Box 981
Columbus, TX 78934
Phone: 409-732-6879
Fax: 409-732-6879

Screening Systems International
P.O. Box 968
Monticello, MS 39654
Phone: 601-587-0522
Fax: 601-587-0524

Seghers Dinamec Inc.
351 Thornton Rd., Ste. 115
Lithia Springs, GA 30057
Phone: 770-739-4205
Fax: 770-944-2236

Selecto, Inc.
1000 Cobb Pl. Blvd., N.W.,
  #370
Kennesaw, GA 30144
Phone: 770-590-1050
Fax: 770-590-0915

Semblex, Inc.
1635 W. Walnut
Springfield, MO 65806-1643
Phone: 417-866-1035
Fax: 417-866-0235

Sentex Systems, Inc.
553 Broad Ave.
Ridgefield, NJ 07657
Phone: 201-945-3694
Fax: 201-941-6064

Sepra Tech, S.A.R.L.
16 Rue de Vigny
F-68000 Colmar, France
Phone: 8923-8416
Fax: 8923-8432

Serck Baker, Inc.
14340 Torrey Chase Blvd.
Houston, TX 77014
Phone: 713-586-8400
Fax: 713-586-9604

Serfilco, Inc.
1777 Shermer Rd.
Northbrook, IL 60062
Phone: 708-998-9300
Fax: 708-998-8929

Sernagiotto S.p.A.
via Torino 144
27045 Casteggio (PV), Italy
Phone: 39-383-83741
Fax: 39-383-83782

Serpentix Conveyor Corp.
9085 Marshall Ct.
Westminster, CO 80030
Phone: 303-430-8427
Fax: 303-430-7337

Servomex Co.
90 Kerry Pl.
Norwood, MA 02062
Phone: 617-769-7710
Fax: 617-769-2834

S&G Enterprises, Inc.
N115 W 1900 Edison Dr.
Germantown, WI 53022
Phone: 414-251-8300
Fax: 414-251-1616

Shannon Chemical Corp.
P.O. Box 376
Malvern, PA 19355
Phone: 610-363-9090
Fax: 610-524-6050

Sierra Silica, Inc.
2260 Athens Ave.
Lincoln, CA 95648
Phone: 916-645-9062
Fax: 916-645-3189

SIHI Pumps, Inc.
P.O. Box 460
Grand Island, NY 14072
Phone: 716-773-6450
Fax: 716-773-2330

Silbrico Corp.
6300 River Rd.
Hodgkins, IL 60525-4257
Phone: 708-354-3350
Fax: 708-354-6698

Simon-Hartley, Ltd.
Stoke-On-Trent
Staffordshire, ST4 7BH
  England
Phone: 0782-202300
Fax: 0782-260534

Sizetec, Inc.
4845 Fulton Dr.
Canton, OH 44718
Phone: 216-492-9682
Fax: 216-492-9041

SLT North America, Inc.
200 S. Trade Center Pkwy.
Conroe, TX 77385
Phone: 713-350-1813
Fax: 409-273-2266

Smith & Loveless, Inc.
14040 Sante Fe Trail Dr.
Lenexa, KS 66215
Phone: 913-888-5201
Fax: 913-888-2173

Solids Technology
  International
Stillorgan Industrial Park
Dublin, Ireland
Phone: 353-1-952229
Fax: 353-1-953874

Solidur Plastics Company
200 Industrial Dr.
Delmont, PA 15626
Phone: 412-468-6868
Fax: 412-468-4044

Solvay Interox
P.O. Box 27328
Houston, TX 77227-7328
Phone: 713-525-6500
Fax: 713-524-9032

Somat Corporation
855 Fox Chase
Coatesville, PA 19320
Phone: 610-384-7000
Fax: 610-380-8500

Spaulding Composites
291 Sam Ridley Pkwy., East
Smyrna, TN 37167
Phone: 615-459-5684
Fax: 615-459-0328

Spencer Turbine Company
600 Day Hill Rd.
Windsor, CT 06095
Phone: 203-688-8361
Fax: 203-688-0098

Sper Chemical Corporation
P.O. Box 5566
Clearwater, FL 33518
Phone: 813-535-9033
Fax: 813-530-0741

Spirac AB
Box 30033
200 61 Malmo, Sweden
Phone: 040-162020
Fax: 040-153650

S.P. Kinney Engineers, Inc.
P.O. Box 445
Carnegie, PA 15106-0445
Phone: 412-276-4600
Fax: 412-276-6890

SRE, Inc.
158 Princeton St.
Nutley, NJ 07110
Phone: 201-661-5192
Fax: 201-808-1242

Stancor, Inc.
515 Fan Hill Rd.
Monroe, CT 06468
Phone: 203-268-7513
Fax: 203-268-7958

Star Systems, Inc.
P.O. Box 518
Timmonsville, SC 29161
Phone: 803-346-3101
Fax: 803-346-3736

Steel Tank Institute
570 Oakwood Rd.
Lake Zurich, IL 60047
Phone: 708-438-8265
Fax: 708-438-4509

Sternson Ltd.
P.O. Box 1540
Brantford, Ontario, Canada
 N3T 5V6
Phone: 519-759-7570
Fax: 519-759-8962

Stevens Water Monitoring
P.O. Box 688
Beaverton, OR 97075
Phone: 503-646-9171
Fax: 503-526-1471

Stiles-Kem, Div. Met-Pro Corp.
3301 Sheridan Rd.
Zion, IL 60099
Phone: 708-746-8334
Fax: 708-746-8341

Stockhausen, Inc.
2408 Doyle St.
Greensboro, NC 27406
Phone: 910-333-3500
Fax: 910-333-3545

Stord, Inc.
309 Regional Rd. South
Greensboro, NC 27409
Phone: 910-668-7727
Fax: 910-668-0537

StormTreat Systems
110 Breed's Hill Rd. #9
Hyannis, MA 02601
Phone: 508-778-4449
Fax: 508-778-4596

Stranco, Inc.
P.O. Box 389
Bradley, IL 60915
Phone: 815-939-1265
Fax: 815-932-0674

SulfaTreat Co.
900 Roosevelt Pkwy.,
  Ste. 610
Chesterfield, MO 63017
Phone: 314-532-2189
Fax: 314-532-2764

Sumitomo Machinery Corp.
4200 Holland Blvd.
Chesapeake, VA 23323
Phone: 804-485-3355
Fax: 804-487-3193

Svedala Industries, Inc.
P.O. Box 15312
York, PA 17405-7312
Phone: 717-843-8671
Fax: 717-845-5154

SWECO Engineering Corp.
7120 New Buffington Rd.
Florence, KY 41042
Phone: 606-727-5147
Fax: 606-727-5122

Sybron Chemicals, Inc.
P.O. Box 66
Birmingham, NJ 08011
Phone: 609-893-1100
Fax: 609-894-8641

Tate Andale, Inc.
1941 Lansdowne Rd.
Baltimore, MD 21227-1789
Phone: 410-247-8700
Fax: 410-247-9672

Taulman Co.
415 E. Paces Ferry
Atlanta, GA 30305-3398
Phone: 404-262-3131
Fax: 404-266-2731

Technical Products, Corp.
1520 High St.
Portsmouth, VA 23704
Phone: 804-399-5009
Fax: 804-397-0914

Techniflo Systems
12300 Perry Hwy
Wexford, PA 15090
Phone: 412-935-0600
Fax: 412-935-0777

Tecnetics Industries, Inc.
1811 Buerkle Rd.
St. Paul, MN 55110
Phone: 612-777-4780
Fax: 612-777-5882

Tellkamp Systems, Inc.
15520 Cornet Ave.
Sante Fe Springs, CA 90670
Phone: 310-802-1621
Fax: 310-802-1303

Tenco Hydro, Inc.
4620 Forest Ave.
Brookfield, IL 60513
Phone: 708-387-0700
Fax: 708-387-0732

Tetko, Inc.
333 South Highland Ave.
Briarcliff Manor, NY 10510
Phone: 914-941-7767
Fax: 914-762-8599

TETRA Technologies, Inc.
P.O. Box 73087
Houston, TX 77273
Phone: 713-367-1983
Fax: 713-364-2270

Thermacon Enviro Systems,
  Inc.
111 W. 40th St.
New York, NY 10018
Phone: 212-704-2111
Fax: 212-704-2089

ThermaFab, Inc.
200 Rich Lex Dr.
Lexington, SC 29072
Phone: 803-794-2543
Fax: 803-796-1602

Thermax, Ltd.
40440 Grand River
Novi, MI 48050
Phone: 313-474-3050
Fax: 313-474-5790

Thetford Systems, Inc.
P.O. Box 1285
Ann Arbor, MI 48106
Phone: 313-769-6000
Fax: 313-761-7842

TIGG Corporation
Box 11661
Pittsburgh, PA 15228
Phone: 412-563-4300
Fax: 412-563-6155

TN Technologies, Inc.
P.O. Box 800
Round Rock, TX 78680-
  0800
Phone: 512-388-9100
Fax: 512-388-9200

Tonka Equipment Company
5115 Industrial St.
Maple Plain, MN 55359
Phone: 612-479-3125
Fax: 612-479-3395

Toray Industries, Inc.
2-2, Nichon-Muro, Chuo-Ku
Tokyo, 103 Japan
Phone: 03-245-5607
Fax: 03-245-5555

Toyobo Co., Ltd.
2-8 Dojima Hama 2-chrome
Kita-ku, Osaka 530, Japan
Phone: 06-348-3360
Fax: 06-348-3332

Transenviro, Inc.
2 Faraday, Ste. B
Irvine, CA 92718
Phone: 714-472-9110
Fax: 714-472-9210

Tri-Mer Corp.
1400 Monroe St., Box 730
Owosso, MI 48867
Phone: 517-723-7838
Fax: 517-723-7844

TriSep Corp.
93 La Patera Ln.
Goleta, CA 93117
Phone: 805-964-8003
Fax: 805-964-1235

Trojan Technologies, Inc.
845 Consortium Ct.
London, Ontario, Canada
 N6E 2S8
Phone: 519-685-6660
Fax: 519-681-8355

Trusty Cook, Inc.
10530 East 59th St.
Indianapolis, IN 46236
Phone: 317-823-6821
Fax: 317-823-6822

TSI Inc.
P.O. Box 64394
St. Paul, MN 55164
Phone: 612-490-2833
Fax: 612-490-3825

Turner Designs
845 W. Maude Ave.
Sunnyvale, CA 94086
Phone: 408-749-0994
Fax: 408-749-0998

Tytronics Inc.
P.O. Box 590
Waltham, MA 02254
Phone: 617-894-0550
Fax: 617-894-9934

Ultra Additives, Inc.
460 Straight St.
Paterson, NJ 07501
Phone: 201-279-1306
Fax: 201-279-0602

Ultrox International
2435 S. Anne St.
Santa Ana, CA 92704-5308
Phone: 714-545-5557
Fax: 714-557-5396

Unifilt Corp.
P.O. Box 389
Zelienople, PA 16063
Phone: 412-452-5008
Fax: 412-452-1044

Unimin Corp.
258 Elm St.
New Canaan, CT 06840
Phone: 203-966-8880
Fax: 203-972-1378

Union Pump Company
4600 W. Dickman Rd.
Battle Creek, MI 49015-1098
Phone: 616-966-4600
Fax: 616-962-3534

United Industries, Inc.
P.O. Box 3838
Baton Rouge, LA 70821
Phone: 504-292-5527
Fax: 504-293-1655

United McGill Corp.
P.O. Box 820
Columbus, OH 43216-0820
Phone: 614-443-0192
Fax: 614-445-8759

Universal Process
  Equipment Co.
P.O. Box 338
Roosevelt, NJ 08555-0338
Phone: 609-443-4545
Fax: 609-259-0644

Unocal Corporation
1511 East Orangethorpe Ave.
Fullerton, CA 92631
Phone: 714-525-9225
Fax: 714-447-5508

U.S. Filter Corp.
P.O. Box 560
Rockford, IL 61105
Phone: 815-877-3041
Fax: 815-877-0172

Val-Matic, Corp.
905 Riverside Dr.
Elmhurst, IL 60126
Phone: 708-941-7600
Fax: 708-941-8042

Vanton Pump & Equipment
  Corp.
201 Sweetland Ave.
Hillside, NJ 07205
Phone: 908-688-4216
Fax: 908-686-9314

Vara International
1201 19th Pl.
Vero Beach, FL 32960
Phone: 407-567-1320
Fax: 407-567-4108

Vaughn Company, Inc.
364 Monte-Elma Rd.
Montesano, WA 98563
Phone: 360-249-4042
Fax: 360-249-6155

Vibra Screw, Inc.
755 755 Union Blvd.
Totowa, NJ 07512
Phone: 201-256-7410
Fax: 201-256-7567

Vikoma International Ltd.
88 Place Rd.
Cowes, Isle of Wight PO31
  7AE, England
Phone: 44-0-983-296021
Fax: 44-0-983-299035

Vincent Corporation
P.O. Box 5747
Tampa, FL 33675
Phone: 813-248-2650
Fax: 813-247-7557

VMI Inc.
1125 N. Maitlen Dr.
Cushing, OK 74023
Phone: 918-225-7000
Fax: 918-225-0333

Vulcan Industries, Inc.
P.O. Box 390
Missouri Valley, IA 51555
Phone: 712-642-2755
Fax: 712-642-4256

Vulcan Peroxidation
  Systems, Inc.
5151 E. Broadway, #600
Tuscon, AZ 85711
Phone: 602-790-8383
Fax: 602-790-8008

Walker Process Equipment
  Co.
840 N. Russell Ave.
Aurora, IL 60506
Phone: 708-892-7921
Fax: 708-892-7951

Wallace & Tiernan, Inc.
25 Main St.
Belleville, NJ 07109-3057
Phone: 201-759-8000
Fax: 201-759-9333

Warren Rupp, Inc.
P.O. Box 1568
Mansfield, OH 44901-1568
Phone: 419-524-8388
Fax: 419-522-7867

Waste Solutions
500 Southland Dr., Ste. 124
Birmingham, AL 35226-
  3711
Phone: 205-823-5231
Fax: 205-823-6820

Waste Tech, Inc.
1931 Industrial Dr.
Libertyville, IL 60048-9738
Phone: 708-367-5150
Fax: 708-367-1787

Waste Water Systems, Inc.
4386 Lilburn Industrial Way
Lilburn, GA 30247
Phone: 770-921-0022
Fax: 770-564-0409

Wastewater Treatment
Systems, Inc.
1235 Elko Dr.
Sunnyvale, CA 94089
Phone: 408-541-8600
Fax: 408-541-8615

Water and Power
Technologies
P.O. Box 27836
Salt Lake City, UT 84127-
0836
Phone: 801-974-5500
Fax: 801-973-9733

Waukesha Fluid Handling
611 Sugar Creek Rd.
Delavan, WI 53115
Phone: 414-728-8267
Fax: 800-252-1088

Weatherly
1100 Spring St., Ste.
800
Atlanta, GA 30309
Phone: 404-873-5030
Fax: 404-873-1303

Weir Pumps, Ltd.
149 Newlands Rd.
Cathcart, Glasgow, G44
4EX Scotland
Phone: 041-637-7141
Fax: 041-637-7358

WesTech Engineering,
Inc.
3605 South West Temple
Salt Lake City, UT
84115
Phone: 801-265-1000
Fax: 801-265-1080

Western Filter Co.
P.O. Box 16323
Denver, CO 80216
Phone: 303-288-2617
Fax: 303-286-9328

Western States Machine
Co.
P.O. Box 327
Hamilton, OH 45012
Phone: 513-863-4758
Fax: 513-863-3846

Western Water
Management, Inc.
1345 Taney N.
Kansas City, MO 64116
Phone: 816-842-0560
Fax: 816-842-6388

Westwood Chemical
Corp.
46 Tower Dr.
Middletown, NY 10940
Phone: 914-692-6721
Fax: 914-695-1906

Wheelabrator, Bio Gro
Division
180 Admiral Cochrane Dr.
Annapolis, MD 21401
Phone: 410-224-0022
Fax: 410-224-0152

Wheelabrator Clean Air
Systems, Inc.
1501 E. Woodfield Rd.,
Ste. 200W
Schaumburg, IL 60173
Phone: 708-706-6900
Fax: 708-706-6996

Wheelabrator, CPC
Engineering
P.O. Box 36, Rte. 20
Sturbridge, MA 01566
Phone: 508-347-7344
Fax: 508-347-7049

Wheelabrator, HPD
Division
55 Shuman Blvd.
Naperville, IL 60563
Phone: 708-357-7330
Fax: 708-717-2247

Wheelabrator, Johnson
Screens
P.O. Box 64118
St. Paul, MN 55164
Phone: 612-636-3900
Fax: 612-636-0889

Wheelabrator, Memtek
Products
28 Cook St.
Billerica, MA 01821
Phone: 508-667-2828
Fax: 508-667-1731

Wheelabrator, Microfloc
Products
Box 36, Route 20
Sturbridge, MA 01566
Phone: 508-347-7344
Fax: 508-347-7049

Wheelabrator, Wiesemann
Products
441 Main St.
Sturbridge, MA 01566
Phone: 508-347-7344
Fax: 508-347-7049

Wil-Flow Inc.
P.O. Box 207
Westport, WA 98595
Phone: 206-268-0405
Fax: 206-268-1925

Wilfley Weber, Inc.
P.O. Box 2330
Denver, CO 80201
Phone: 303-779-1777
Fax: 303-779-1277

William R. Perrin, Inc.
432 Monarch Ave.
Ajax, Ontario, Canada
  L1S 2G7
Phone: 416-683-9400
Fax: 416-427-2361

W.L. Gore & Associates,
  Inc.
P.O. Box 1100
Elkton, MD 21922
Phone: 410-392-3300
Fax: 410-398-6624

WRc Process Engineering
Aynho Rd., Adderbury,
  Banbury
Oxan, OX17 3NL England
Phone: 0295-812282
Fax: 0295-812283

WTC Industries, Inc.
14405 21st Ave. North
Minneapolis, MN 55447
Phone: 612-473-1625
Fax: 612-473-1712

Wyssmont Company, Inc.
P.O. Box 1397
Fort Lee, NJ 07024
Phone: 201-947-4600
Fax: 201-947-0324

Yeomans Chicago Corp.
1999 North Ruby St.
Melrose Park, IL 60160
Phone: 708-344-9600
Fax: 708-681-4432

York Energy Conservation
55 W. Beaver Creek
Richmond Hill, Ontario,
  Canada L4B 1K5
Phone: 416-764-3232
Fax: 416-764-8325

Zenon Environmental, Inc.
845 Harrington Ct.,
  Burlington
Ontario, Canada L7N 3P3
Phone: 905-639-6320
Fax: 905-639-1812

Zeol Division-Munters
  Corp.
79 Monroe St.
Amesbury, MA 01913
Phone: 508-388-2666
Fax: 508-388-5553

Zimpro Environmental, Inc.
301 W. Military Rd.
Rothschild, WI 54474
Phone: 715-359-7211
Fax: 715-355-3219

Z Polymetron
405 Barclay Blvd.
Lincolnshire, IL 60069
Phone: 708-634-0000
Fax: 708-634-1371

Zurn Industries, Inc.
P.O. Box 668
Woodbridge, NJ 07095-0668
Phone: 908-634-1761
Fax: 908-634-0798

Printed and bound by CPI Group (UK) Ltd, Croydon, CR0 4YY

21/10/2024

01777049-0001